AF553413

Environmental Conservation Issues and Concepts

Environmental Conservation Issues and Concepts

Tanuj Sharma

RANDOM PUBLICATIONS
NEW DELHI (INDIA)

Environmental Conservation Issues and Concepts

ISBN 978-93-5111-471-0

Published in 2015 in India by

RANDOM PUBLICATIONS

4376-A/4B, Gali Murari Lal, Ansari Road
New Delhi-110 002
Phone : +9111-43580356, 011-23289044, 011-43142548
e-mail: sales@randompublications.com,
info@randompublications.com, randomexports@gmail.com

Reprinted 2022

Type Setting by : Friends Media, Delhi-110089
Digitally Printed at : Replika Press Pvt. Ltd.

Preface

Environmental conservation is a practice of protecting and conserving the environment, on individual, organisational or governmental level, for the benefit of the natural environment and humans. Protection of the environment is needed from various human activities. Waste, pollution, loss of biodiversity, introduction of invasive species, release of genetically modified organisms and toxics are some of the issues relating to environmental protection.

Due to the pressures of population and our technology the biophysical environment is being degraded, sometimes permanently. Many of the earth's resources are especially vulnerable because they are influenced by human impacts across many countries. As a result of this, many attempts are made by countries to develop agreements that are signed by multiple governments to prevent damage or manage the impacts of human activity on natural resources. This can include agreements that impact factors such as climate, oceans, rivers and air pollution. These international environmental agreements are sometimes legally binding documents that have legal implications when they are not followed and, at other times, are more agreements in principle or are for use as codes of conduct. Some of the most well-known multinational agreements include: the Kyoto Protocol, Vienna Convention on the Protection of the Ozone Layer and Rio Declaration on Development and Environment.

This book presents up-to-date information about global environmental issues and emphasises on the importance of conserving the natural resources. The international policies and programmes for environmental conservation is described in detail. It will be a helpful resource for students, research scholars, professors, scientists as well as for policy makers in achieving the goal of better environment.

Author

Contents

Preface v

1. Importance of Environmental Protection **1**

Understanding the Natural Environment 1

Challenges to Natural Environment 11

Approaches to Environmental Protection 12

Actions of Government 14

2. History of Environmental and Biodiversity Conservation **21**

Conservation Movement 21

History of Conservation Ideas 22

Areas of Concern 23

History of Biodiversity Conservation 25

Natural Variations in Ecology and Biodiversity 30

Imprints of Environmental Change on Biodiversity 30

Ecological Footprint of the Human Race 33

3. Community-Based Environmental Protection **36**

Community-based Environmental Protection 36

Goal-Setting and Developing an Organization 38

Assessing the Conditions of Local Ecosystems 43

Role of Government Agencies 49

Role of Non-Governmental Organizations 49

Role of Local Resources 49

Links Between Ecosystems and the Local Economy 50

Ecosystems and Quality of Life 51

4. Land Use Management and Soil Conservation **53**

MDGs and Sustainable Land Use Management and Soil Conservation 53

UNEP's Land-related Activities 55

World Soil Charter 56

Global Environmental Outlook 56

Key Issues 58

Challenges 61

UNEP's Goals to Sustainable Development 62

Strategies and Intended Action 64

Integrated Water Resource Management 69

Conserving Below-ground Biodiversity 70

Agri-food Production and Consumption 71

Global Land Cover Monitoring and Assessment 72

Policy Development and Implementation 72

Land Degradation Control Initiative by UNEP 74

GEF Operational Programme #15 75

Improved Science-policy Interaction and Knowledge Systems 76

Implementation of the UN Convention to Combat Desertification and Support to Africa 78

UNEP and Civil Society 79

The Environment Initiative of NEPAD 79

Mechanisms for Implementation and Resources 80

5. Globalization and Environmental Protection **83**

Effects of Globalisation on the Environment 84

Environmental Effects on Globalisation 88

Global Environmental Policies 91

Global Environmental Governance 93

Reforms in Environmental and Economic Governance 94

Governance Alternatives 96

Global Environmental Information Clearing-House 99

Technological Advances and Environmental Conservation 100
Responses to Global-scale Environmental Problems 101

6. Forest Land Conservation 104
Establishment of Protected Areas 104
Political and Economic Costs of Protected Areas 106
The Need to Consider Alternatives 109
Challenging of Forest Protection 112
Policy Implications 120

7. Rights-based Approaches to Environmental Conservation 123
Understanding Rights 123
Recognition of Environment 127
Relationships between Rights and Conservation 127
Concepts and Components for RBAs to Conservation 130
Accountability in RBAs to Conservation 131
Conservation, Rights, and Governance 134
Rights Relevant for RBAs to Conservation 136
Conservation Approaches 141
Adopting RBAs to Conservation 142
Developing and Implementing RBA to Cnservation 145
Strategies to Integrate Rights Considerations into Practice 147

8. Conservation and Human Rights 151
Hard Law Provisions 152
Soft Laws 160
Funding Agencies 162
International Conservation NGOs 168

9. World Conservation Strategy 176
Living Resource Conservation for Sustainable Development 178
Objectives of Conservation and Requirements for their Achievement 182
Priorities for National Action 208
Improving the Capacity to Manage 223

Training and Research 228
Participation and Education 233
Conservation-based Rural Development 238
Priorities for International Action 242
Global Programme for the Protection of Genetic Resource Areas 251
The Global Commons 256
Regional Strategies for International River Basin and Sea 261
Towards Sustainable Development 266

Bibliography **270**

Index **273**

1

Importance of Environmental Protection

Understanding the Natural Environment

The natural environment encompasses all living and non-living things occurring naturally on Earth or some region thereof. It is an environment that encompasses the interaction of all living species. The concept of the natural environment can be distinguished by components:

Complete ecological units that function as natural systems without massive human intervention, including all vegetation, microorganisms, soil, rocks, atmosphere, and natural phenomena that occur within their boundaries.

Universal natural resources and physical phenomena that lack clear-cut boundaries, such as air, water, and climate, as well as energy, radiation, electric charge, and magnetism, not originating from human activity.

The natural environment is contrasted with the built environment, which comprises the areas and components that are strongly influenced by humans. A geographical area is regarded as a natural environment.

It is difficult to find absolutely natural environments, and it is common that the naturalness varies in a continuum, from ideally 100% natural in one extreme to 0% natural in the other. More precisely, we can consider the different aspects or components of an environment, and see that their degree of naturalness is not uniform. If, for instance, we take an agricultural field, and consider the mineralogic composition and the structure of its soil, we will find that whereas the first is quite similar to that of an undisturbed forest soil, the structure is quite different.

Natural environment is often used as a synonym for habitat. For instance, when we say that the natural environment of giraffes is the savanna.

Composition

Earth science generally recognizes 4 spheres, the lithosphere, the hydrosphere, the atmosphere, and the biosphere as correspondent to rocks, water, air, and life. Some scientists include, as part of the spheres of the Earth, the cryosphere (corresponding to ice) as a distinct portion of the hydrosphere, as well as the pedosphere (corresponding to soil) as an active and intermixed sphere. Earth science (also known as geoscience, the geosciences or the Earth Sciences), is an all-embracing term for the sciences related to the planet Earth. There are four major disciplines in earth sciences, namely geography, geology, geophysics and geodesy. These major disciplines use physics, chemistry, biology, chronology and mathematics to build a qualitative and quantitative understanding of the principal areas or spheres of the Earth system.

The Earth's crust, or lithosphere, is the outermost solid surface of the planet and is chemically and mechanically different from underlying mantle. It has been generated largely by igneous processes in which magma (molten rock) cools and solidifies to form solid rock. Beneath the lithosphere lies the mantle which is heated by the decay of radioactive elements. The mantle though solid is in a state of rheic convection. This convection process causes the lithospheric plates to move, albeit slowly. The resulting process is known as plate tectonics. Volcanoes result primarily from the melting of subducted crust material or of rising mantle at mid-ocean ridges and mantle plumes.

Water on Earth

An ocean is a major body of saline water, and a component of the hydrosphere. Approximately 71% of the Earth's surface (an area of some 362 million square kilometers) is covered by ocean, a continuous body of water that is customarily divided into several principal oceans and smaller seas. More than half of this area is over 3,000 meters (9,800 ft) deep. Average oceanic salinity is around 35 parts per thousand (ppt) (3.5%), and nearly all seawater has a salinity in the range of 30 to 38 ppt. Though generally recognized as several 'separate' oceans, these waters comprise one global, interconnected body of salt water often referred to as the World Ocean or global ocean. This concept of a global ocean as a continuous body of water with relatively free interchange among its parts is of fundamental importance to oceanography. The major oceanic divisions are defined in part by the continents, various archipelagos, and other criteria: these divisions are (in descending order of size) the Pacific Ocean, the Atlantic Ocean, the Indian Ocean, the Southern Ocean and the Arctic Ocean.

Rivers

A river is a natural watercourse, usually freshwater, flowing toward an ocean, a lake, a sea or another river. In a few cases, a river simply flows into the ground or dries up completely before reaching another body of water. Small rivers may also be termed by several other names, including stream, creek and brook. In the United States a river is generally classified as a watercourse more than 60 feet (18 metres) wide. The water in a river is usually in a channel, made up of a stream bed between banks. In larger rivers there is also a wider floodplain shaped by waters over-topping the channel. Flood plains may be very wide in relation to the size of the river channel. Rivers are a part of the hydrological cycle. Water within a river is generally collected from precipitation through surface runoff, groundwater recharge, springs, and the release of water stored in glaciers and snowpacks.

Streams

A stream is a flowing body of water with a current, confined within a bed and stream banks. Streams play an important corridor role in connecting fragmented habitats and thus in conserving biodiversity. The study of streams and waterways in general is known as surface hydrology. Types of streams include creeks, tributaries, which do not reach an ocean and connect with another stream or river, brooks, which are typically small streams and sometimes sourced from a spring or seep and tidal inlets.

Lakes

A lake (from Latin lacus) is a terrain feature, a body of water that is localized to the bottom of basin. A body of water is considered a lake when it is inland, is not part of a ocean, is larger and deeper than a pond, and is fed by a river.

Natural lakes on Earth are generally found in mountainous areas, rift zones, and areas with ongoing or recent glaciation. Other lakes are found in endorheic basins or along the courses of mature rivers. In some parts of the world, there are many lakes because of chaotic drainage patterns left over from the last Ice Age. All lakes are temporary over geologic time scales, as they will slowly fill in with sediments or spill out of the basin containing them.

Ponds

A pond is a body of standing water, either natural or man-made, that is usually smaller than a lake. A wide variety of man-made bodies of water are classified as ponds, including water gardens designed for aesthetic ornamentation, fish ponds designed for commercial fish breeding, and solar ponds designed to store thermal energy. Ponds and lakes are distinguished from streams via current speed. While currents in streams are easily observed, ponds and lakes possess thermally driven micro-currents and moderate wind driven currents. These features distinguish a pond from many other aquatic terrain features, such as stream pools and tide pools.

Atmosphere, Climate and Weather

The atmosphere of the Earth serves as a key factor in sustaining the planetary ecosystem. The thin layer of gases that envelops the Earth is held in place by the planet's gravity. Dry air consists of 78% nitrogen, 21% oxygen, 1% argon and other inert gases, such as carbon dioxide.

The remaining gases are often referred to as trace gases, among which are the greenhouse gases such as water vapor, carbon dioxide, methane, nitrous oxide, and ozone. Filtered air includes trace amounts of many other chemical compounds.

Air also contains a variable amount of water vapor and suspensions of water droplets and ice crystals seen as clouds. Many natural substances may be present in tiny amounts in an unfiltered air sample, including dust, pollen and spores, sea spray, volcanic ash, and meteoroids. Various industrial pollutants also may be present, such as chlorine (elementary or in compounds), fluorine compounds, elemental mercury, and sulphur compounds such as sulphur dioxide [SO_2].

The ozone layer of the Earth's atmosphere plays an important role in depleting the amount of ultraviolet (UV) radiation that reaches the surface. As DNA is readily damaged by UV light, this serves to protect life at the surface. The atmosphere also retains heat during the night, thereby reducing the daily temperature extremes.

Principal Layers

Earth's atmosphere can be divided into five main layers. These layers are mainly determined by whether temperature increases or decreases with altitude. From highest to lowest, these layers are:

- Exosphere: The outermost layer of Earth's atmosphere extends from the exobase upward, mainly composed of hydrogen and helium.
- Thermosphere: The top of the thermosphere is the bottom of the exosphere, called the exobase. Its height varies with solar activity and ranges from about 350–800 km (220–500 mi; 1,100,000–2,600,000 ft). The International Space Station orbits in this layer, between 320 and 380 km (200 and 240 mi).
- Mesosphere: The mesosphere extends from the stratopause to 80–85 km (50–53 mi; 260,000–280,000 ft). It is the layer where most meteors burn up upon entering the atmosphere.
- Stratosphere: The stratosphere extends from the tropopause to about 51 km (32 mi; 170,000 ft). The stratopause, which is the boundary between the stratosphere and mesosphere, typically is at 50 to 55 km (31 to 34 mi; 160,000 to 180,000 ft).
- Troposphere: The troposphere begins at the surface and extends to between 7 km (23,000 ft) at the poles and 17 km (56,000 ft) at the equator, with some variation due to weather. The troposphere is mostly heated by transfer of energy from the surface, so on average the lowest part of the troposphere is warmest and temperature decreases with altitude. The tropopause is the boundary between the troposphere and stratosphere.

Other layers

Within the five principal layers determined by temperature are several layers determined by other properties.

- The ozone layer is contained within the stratosphere. It is mainly located in the lower portion of the stratosphere from about 15–35 km (9.3–22 mi; 49,000–110,000 ft), though the thickness varies seasonally and geographically. About 90% of the ozone in our atmosphere is contained in the stratosphere.
- The ionosphere, the part of the atmosphere that is ionized by solar radiation, stretches from 50 to 1,000 km (31 to 620 mi; 160,000 to

3,300,000 ft) and typically overlaps both the exosphere and the thermosphere. It forms the inner edge of the magnetosphere.

— The homosphere and heterosphere: The homosphere includes the troposphere, stratosphere, and mesosphere. The upper part of the heterosphere is composed almost completely of hydrogen, the lightest element.

— The planetary boundary layer is the part of the troposphere that is nearest the Earth's surface and is directly affected by it, mainly through turbulent diffusion.

Effects of global warming

The potential dangers of global warming are being increasingly studied by a wide global consortium of scientists. These scientists are increasingly concerned about the potential long-term effects of global warming on our natural environment and on the planet. Of particular concern is how climate change and global warming caused by anthropogenic, or human-made releases of greenhouse gases, most notably carbon dioxide, can act interactively, and have adverse effects upon the planet, its natural environment and humans' existence. Efforts have been increasingly focused on the mitigation of greenhouse gases that are causing climatic changes, on developing adaptative strategies to global warming, to assist humans, animal and plant species, ecosystems, regions and nations in adjusting to the effects of global warming. Some examples of recent collaboration to address climate change and global warming include:

— The United Nations Framework Convention Treaty and convention on Climate Change, to stabilize greenhouse gas concentrations in the atmosphere at a level that would prevent dangerous anthropogenic interference with the climate system.

— The Kyoto Protocol, which is the protocol to the international Framework Convention on Climate Change treaty, again with the objective of reducing greenhouse gases in an effort to prevent anthropogenic climate change.

— The Western Climate Initiative, to identify, evaluate, and implement collective and cooperative ways to reduce greenhouse gases in the region, focusing on a market-based cap-and-trade system.

A significantly profound challenge is to identify the natural environmental dynamics in contrast to environmental changes not within natural variances.

A common solution is to adapt a static view neglecting natural variances to exist. Methodologically, this view could be defended when looking at processes which change slowly and short time series, while the problem arrives when fast processes turns essential in the object of the study.

Climate

Climate encompasses the statistics of temperature, humidity, atmospheric pressure, wind, rainfall, atmospheric particle count and numerous other meteorological elements in a given region over long periods of time. Climate can be contrasted to weather, which is the present condition of these same elements over periods up to two weeks.

Climates can be classified according to the average and typical ranges of different variables, most commonly temperature and precipitation. The most commonly used classification scheme is the one originally developed by Wladimir Köppen. The Thornthwaite system, in use since 1948, incorporates evapotranspiration in addition to temperature and precipitation information and is used in studying animal species diversity and potential impacts of climate changes.

Weather

Weather is a set of all the phenomena occurring in a given atmospheric area at a given time. Most weather phenomena occur in the troposphere, just below the stratosphere. Weather refers, generally, to day-to-day temperature and precipitation activity, whereas climate is the term for the average atmospheric conditions over longer periods of time. When used without qualification, "weather" is understood to be the weather of Earth.

Weather occurs due to density (temperature and moisture) differences between one place and another. These differences can occur due to the sun angle at any particular spot, which varies by latitude from the tropics. The strong temperature contrast between polar and tropical air gives rise to the jet stream. Weather systems in the mid-latitudes, such as extratropical cyclones, are caused by instabilities of the jet stream flow. Because the Earth's axis is tilted relative to its orbital plane, sunlight is incident at different angles at different times of the year. On the Earth's surface, temperatures usually range ±40 °C (100 °F to -40 °F) annually. Over thousands of years, changes in the Earth's orbit have affected the amount and distribution of solar energy received by the Earth and influence long-term climate

Surface temperature differences in turn cause pressure differences. Higher altitudes are cooler than lower altitudes due to differences in compressional heating. Weather forecasting is the application of science and technology to predict the state of the atmosphere for a future time and a given location. The atmosphere is a chaotic system, and small changes to one part of the system can grow to have large effects on the system as a whole. Human attempts to control the weather have occurred throughout human history, and there is evidence that human activity such as agriculture and industry has inadvertently modified weather patterns.

Life

Evidence suggest that life on Earth has existed for about 3.7 billion years. All known life forms share fundamental molecular mechanisms, and based on these observations, theories on the origin of life attempt to find a mechanism explaining the formation of a primordial single cell organism from which all life originates. There are many different hypotheses regarding the path that might have been taken from simple organic molecules via pre-cellular life to protocells and metabolism.

Although there is no universal agreement on the definition of life, scientists generally accept that the biological manifestation of life is characterized by organization, metabolism, growth, adaptation, response to stimuli and reproduction. Life may also be said to be simply the characteristic state of organisms. In biology, the science of living organisms, "life" is the condition which distinguishes active organisms from inorganic matter, including the capacity for growth, functional activity and the continual change preceding death.

A diverse array of living organisms (life forms) can be found in the biosphere on Earth, and properties common to these organisms—plants, animals, fungi, protists, archaea, and bacteria—are a carbon- and water-based cellular form with complex organization and heritable genetic information. Living organisms undergo metabolism, maintain homeostasis, possess a capacity to grow, respond to stimuli, reproduce and, through natural selection, adapt to their environment in successive generations. More complex living organisms can communicate through various means.

Ecosystems

An ecosystem (also called as environment) is a natural unit consisting of all plants, animals and micro-organisms (biotic factors) in an area functioning together with all of the non-living physical (abiotic) factors of the environment.

Central to the ecosystem concept is the idea that living organisms are continually engaged in a highly interrelated set of relationships with every other element constituting the environment in which they exist. Eugene Odum, one of the founders of the science of ecology, stated: "Any unit that includes all of the organisms (ie: the "community") in a given area interacting with the physical environment so that a flow of energy leads to clearly defined trophic structure, biotic diversity, and material cycles (i.e.: exchange of materials between living and nonliving parts) within the system is an ecosystem."

The human ecosystem concept is then grounded in the deconstruction of the human/nature dichotomy, and the emergent premise that all species are ecologically integrated with each other, as well as with the abiotic constituents of their biotope.

A greater number or variety of species or biological diversity of an ecosystem may contribute to greater resilience of an ecosystem, because there are more species present at a location to respond to change and thus "absorb" or reduce its effects. This reduces the effect before the ecosystem's structure is fundamentally changed to a different state. This is not universally the case and there is no proven relationship between the species diversity of an ecosystem and its ability to provide goods and services on a sustainable level.

The term ecosystem can also pertain to human-made environments, such as human ecosystems and human-influenced ecosystems, and can describe any situation where there is relationship between living organisms and their environment. Fewer areas on the surface of the earth today exist free from human contact, although some genuine wilderness areas continue to exist without any forms of human intervention.

Biomes

Biomes are terminologically similar to the concept of ecosystems, and are climatically and geographically defined areas of ecologically similar climatic conditions on the Earth, such as communities of plants, animals, and soil organisms, often referred to as ecosystems. Biomes are defined on the basis of factors such as plant structures (such as trees, shrubs, and grasses), leaf types (such as broadleaf and needleleaf), plant spacing (forest, woodland, savanna), and climate. Unlike ecozones, biomes are not defined by genetic, taxonomic, or historical similarities. Biomes are often identified with particular patterns of ecological succession and climax vegetation.

Biogeochemical cycles

Global biogeochemical cycles are critical to life, most notably those of water, oxygen, carbon, nitrogen and phosphorus.

- The nitrogen cycle is the transformation of nitrogen and nitrogen-containing compounds in nature. It is a cycle which includes gaseous components.
- The water cycle, is the continuous movement of water on, above, and below the surface of the Earth. Water can change states among liquid, vapor, and ice at various places in the water cycle. Although the balance of water on Earth remains fairly constant over time, individual water molecules can come and go.
- The carbon cycle is the biogeochemical cycle by which carbon is exchanged among the biosphere, pedosphere, geosphere, hydrosphere, and atmosphere of the Earth.
- The oxygen cycle is the movement of oxygen within and between its three main reservoirs: the atmosphere, the biosphere, and the lithosphere. The main driving factor of the oxygen cycle is photosynthesis, which is responsible for the modern Earth's atmospheric composition and life.
- The phosphorus cycle is the movement of phosphorus through the lithosphere, hydrosphere, and biosphere. The atmosphere does not play a significant role in the movements of phosphorus, because phosphorus and phosphorus compounds are usually solids at the typical ranges of temperature and pressure found on Earth.

Wilderness

Wilderness is generally defined as a natural environment on Earth that has not been significantly modified by human activity. The WILD Foundation goes into more detail, defining wilderness as: "The most intact, undisturbed wild natural areas left on our planet - those last truly wild places that humans do not control and have not developed with roads, pipelines or other industrial infrastructure." Wilderness areas and protected parks are considered important for the survival of certain species, ecological studies, conservation, solitude, and recreation. Wilderness is deeply valued for cultural, spiritual, moral, and aesthetic reasons. Some nature writers believe wilderness areas are vital for the human spirit and creativity.

The word, "wilderness", derives from the notion of wildness; in other words that which is not controllable by humans. The word's etymology is from the Old English wildeornes, which in turn derives from wildeor meaning wild beast. From this point of view, it is the wildness of a place that makes it a wilderness. The mere presence or activity of people does not disqualify an area from being "wilderness." Many ecosystems that are, or have been, inhabited or influenced by activities of people may still be considered "wild." This way of looking at wilderness includes areas within which natural processes operate without very noticeable human interference.

Wildlife includes all non-domesticated plants, animals and other organisms. Domesticating wild plant and animal species for human benefit has occurred many times all over the planet, and has a major impact on the environment, both positive and negative. Wildlife can be found in all ecosystems. Deserts, rain forests, plains, and other areas—including the most developed urban sites—all have distinct forms of wildlife. While the term in popular culture usually refers to animals that are untouched by human factors, most scientists agree that wildlife around the world is impacted by human activities.

Challenges to Natural Environment

It is the common understanding of natural environment that underlies environmentalism — a broad political, social, and philosophical movement that advocates various actions and policies in the interest of protecting what nature remains in the natural environment, or restoring or expanding the role of nature in this environment. While true wilderness is increasingly rare, wild nature (e.g., unmanaged forests, uncultivated grasslands, wildlife, wildflowers) can be found in many locations previously inhabited by humans.

Goals commonly expressed by environmental scientists include:

- Reduction and clean up of pollution, with future goals of zero pollution;
- Cleanly converting non-recyclable materials into energy through direct combustion or after conversion into secondary fuels;

— Reducing societal consumption of non-renewable fuels;
— Development of alternative, green, low-carbon or renewable energy sources;
— Conservation and sustainable use of scarce resources such as water, land, and air;
— Protection of representative or unique or pristine ecosystems;
— Preservation of threatened and endangered species extinction;

The establishment of nature and biosphere reserves under various types of protection; and, most generally, the protection of biodiversity and ecosystems upon which all human and other life on earth depends.

Very large development projects - megaprojects - pose special instructions and risks to the natural environments. Major dams and power plants are cases in point. The challenge to the environment from such projects is growing because more and bigger megaprojects are being built, in developed and developing nations alike.

Approaches to Environmental Protection

Environmental protection is a practice of protecting the natural environment on individual, organizational or governmental levels, for the benefit of the natural environment and humans. Due to the pressures of population and technology, the biophysical environment is being degraded, sometimes permanently. This has been recognized, and governments have begun placing restraints on activities that cause environmental degradation. Since the 1960s, activity of environmental movements has created awareness of the various environmental issues. There is no agreement on the extent of the environmental impact of human activity, and protection measures are occasionally criticized.

Academic institutions now offer courses, such as environmental studies, environmental management and environmental engineering, that teach the history and methods of environment protection. Protection of the environment is needed due to various human activities. Waste production, air pollution, and loss of biodiversity (resulting from the introduction of invasive species and species extinction) are some of the issues related to environmental protection.

Environmental protection is influenced by three interwoven factors: environmental legislation, ethics and education. Each of these factors plays its part in influencing national-level environmental decisions and personal-level environmental values and behaviors. For environmental protection to become a reality, it is important for societies to develop each of these areas that, together, will inform and drive environmental decisions.

Voluntary Environmental Agreements

In industrialized countries, voluntary environmental agreements often provide a platform for companies to be recognized for moving beyond the minimum regulatory standards and, thus, support the development of best environmental practice. In developing countries, such as throughout Latin America, these agreements are more commonly used to remedy significant levels of non-compliance with mandatory regulation. The challenges that exist with these agreements lie in establishing baseline data, targets, monitoring and reporting. Due to the difficulties inherent in evaluating effectiveness, their use is often questioned and, indeed, the environment may well be adversely affected as a result. The key advantage of their use in developing countries is that their use helps to build environmental management capacity.

Ecosystems Approach

An ecosystems approach to resource management and environmental protection aims to consider the complex interrelationships of an entire ecosystem in decision making rather than simply responding to specific issues and challenges. Ideally the decision-making processes under such an approach would be a collaborative approach to planning and decision making that involves a broad range of stakeholders across all relevant governmental departments, as well as representatives of industry, environmental groups and community. This approach ideally supports a better exchange of information, development of conflict-resolution strategies and improved regional conservation.

International Environmental Agreements

Many of the earth's resources are especially vulnerable because they are influenced by human impacts across many countries. As a result of this, many attempts are made by countries to develop agreements that are signed by multiple governments to prevent damage or manage the impacts of human activity on natural resources. This can include agreements that impact factors such as climate, oceans, rivers and air pollution. These international environmental agreements are sometimes legally binding documents that have legal implications when they are not followed and, at other times, are more agreements in principle or are for use as codes of conduct. These agreements have a long history with some multinational agreements being in place from as early as 1910 in Europe, America and Africa. Some of the most well-known multinational agreements include: the Kyoto Protocol, Vienna Convention on the Protection of the Ozone Layer and Rio Declaration on Development and Environment.

Actions of Government

Discussion concerning environmental protection often focuses on the role of government, legislation and law enforcement. However, in its broadest sense, environmental protection may be seen to be the responsibility of all people and not simply that of government. Decisions that impact the environment will ideally involve a broad range of stakeholders, including industry, indigenous groups, environmental group and community representatives. Gradually, environmental decision-making processes are evolving to reflect this broad base of stakeholders and are becoming more collaborative in many countries.

Many constitutions acknowledge the fundamental right to environmental protection, and many international treaties acknowledge the right to live in a healthy environment. Also, many countries have organizations and agencies devoted to environmental protection. There are international environmental protection organizations, as the United Nations Environment Programme.

Although environmental protection is not simply the responsibility of government agencies, most people view these agencies as being of prime importance in establishing and maintaining basic standards that protect both the environment and the people interacting with it.

Africa

Tanzania is recognised as having some of the greatest biodiversity of any African country. Almost 40% of the land has been established into a network of protected areas, including several national parks. The concerns for the natural environment include damage to ecosystems and loss of habitat resulting from population growth, expansion of subsistence agriculture, pollution, timber extraction and significant use of timber as fuel.

Environmental protection in Tanzania began during the German occupation of East Africa (1884-1919)—colonial conservation laws for the protection of game and forests were enacted, whereby restrictions were placed upon traditional indigenous activities, such as hunting, firewood collecting and cattle grazing. In 1948, Serengeti was officially established as the first national park for wild cats in East Africa. Since 1983, there has been a more broad-reaching effort to manage environmental issues at a national level, through the establishment of the National Environment Management Council (NEMC) and the development of an environmental act.

The Division of the Environment is the main government body that oversees protection. It does this through formulation of policy, coordinating and monitoring environmental issues, environmental planning and policy-oriented environmental research.The National Environment Management Council (NEMC) is an institution that was initiated when the National Environment Management Act was first introduced in 1983. This council has the role to advise governments and the international community on a range of environmental issues. The NEMC has the following purposes: provide technical advice; coordinate technical activities; develop enforcement guidelines and procedures; assess, monitor and evaluate activities that impact the environment; promote and assist environmental information and communication; and seek advancement of scientific knowledge.

The National Environment Policy of 1997 acts as a framework for environmental decision making in Tanzania. The policy objectives are to:

— Ensure sustainable and equitable use of resources without degrading the environment or risking health or safety
— Prevent and control degradation of land, water, vegetation and air
— Conserve and enhance natural and man-made heritage, including biological diversity of unique ecosystems
— Improve condition and productivity of degraded areas
— Raise awareness and understanding of the link between environment and development
— Promote individual and community participation
— Promote international cooperation

Tanzania is a signatory to a significant number of international conventions, including the Rio Declaration on Development and Environment 1992 and the Convention on Biological Diversity 1996. The Environmental Management Act, 2004, is the first comprehensive legal and institutional framework to guide environmental-management decisions. The policy tools that are parts of the act includes the use of: environmental-impact assessments, strategics environmentals assessments and taxation on pollution for specific industries and products. The effectiveness of shifing of this act will only become clear over time as concerns regarding its implementation become apparent based on the fact that, historically, there has been a lack of capacity to enforce environmental laws and a lack of working tools to bring environmental-protection objectives into practice.

Asia

Formal environmental protection in China was first stimulated by the 1972 United Nations Conference on the Human Environment, held in Stockholm, Sweden. Following this, China began establishing environmental protection agencies and putting controls on some of its industrial waste. China was one of the first developing countries to implement a sustainable development strategy. In 1983 the State Council announced that environmental protection would be one of China's basic national policies and in 1984 the National Environmental Protection Agency (NEPA) was established. Following severe flooding of the Yangtze River basin in 1998, NEPA was upgraded to the State Environmental Protection Agency (SEPA) meaning that environmental protection was now being implemented at a ministerial level. In 2008, SEPA became known by its current name of Ministry of Environmental Protection of the People's Republic of China (MEP).

Environmental pollution and ecological degradation has resulted in economic losses for China. In 2005, economic losses (mainly from air pollution) were calculated at 7.7% of China's GDP. This grew to 10.3% by 2002 and the economic loss from water pollution (6.1%) began to exceed that caused by air pollution. China has been one of the top performing countries in terms of GDP growth (9.64% in the past ten years). However, the high economic growth has put immense pressure on its environment and the environmental challenges that China faces are greater than most countries. In 2010 China was ranked 121st out of 163 countries on the Environmental Performance Index.

China has taken initiatives to increase its protection of the environment and combat environmental degradation:

— China's investment in renewable energy grew 18% in 2007 to $15.6 billion, accounting for ~10% of the global investment in this area;).

— In 2008, spending on the environment was 1.49% of GDP, up 3.4 times from 2000;

— The discharge of COD (carbon monoxide) and SO2 (sulfur dioxide) decreased by 6.61% and 8.95% in 2008 compared with that in 2005;

— China's protected nature reserves have increased substantially. In 1978 there were only 34 compared with 2,538 in 2010. The protected nature reserve system now occupies 15.5% of the country; this is higher than the world average.

Rapid growth in GDP has been China's main goal during the past three decades with a dominant development model of inefficient resource use and high pollution to achieve high GDP. For China to develop sustainably, environmental protection should be treated as an integral part of its economic policies.

Quote from Shengxian Zhou, head of MEP (2009): "Good economic policy is good environmental policy and the nature of environmental problem is the economic structure, production form and develop model."

European Union

Environmental protection has become an important task for the institutions of the European Community after the Maastricht Treaty for the European Union ratification by all Member States. The EU is already very active in the field of environmental policy with important directives like those on environmental impact assessment and on the access to environmental information for citizens in the Member States.

Latin America

The United Nations Environment Programme (UNEP) has identified 17 megadiverse countries. The list includes six Latin American countries: Brazil, Colombia, Ecuador, Mexico, Peru and Venezuela. Mexico and Brazil stand out among the rest because they have the largest area, population and number of species. These countries represent a major concern for environmental protection because they have high rates of deforestation, ecosystems loss, pollution, and population growth.

The Secretariat of Environment and Natural Resources (Secretaría del Medio Ambiente y Recursos Naturales, SEMARNAT) is Mexico's environment ministry. The Ministry is responsible for addressing the following issues:

— Promote the protection, restoration and conservation of ecosystems, natural resources, goods and environmental services, and to facilitate their use and sustainable development.
— Develop and implement a national policy on natural resources
— Promote environmental management within the national territory, in coordination with all levels of government and the private sector.
— Evaluate and provide determination to the environmental impact statements for development projects and prevention of ecological damage
— Implement national policies on climate change and protection of the ozone layer.
— Direct work and studies on national meteorological, climatological, hydrological, and geohydrological systems, and participate in international conventions on these subjects.
— Regulate and monitor the conservation of waterways

In November 2000 there were 127 protected areas; currently there are 174, covering an area of 25,384,818 hectares, increasing federally protected areas from 8.6% to 12.85% its land area.

Oceania

In 2008 there was 98,487,116 ha of terrestrial protected area, covering 12.8% of the land area of Australia. The 2002 figures of 10.1% of terrestrial area and 64,615,554 ha of protected marine area were found to poorly represent about half of Australia's 85 bioregions.

Environmental protection in Australia could be seen as starting with the formation of the first National Park, Royal National Park, in 1879. More progressive environmental protection had it start in the 1960s and 1970s with major international programs such as the United Nations Conference on the Human Environment in 1972, the Environment Committee of the OECD in 1970, and the United Nations Environment Programme of 1972. These events laid the foundations by increasing public awareness and support for regulation. State environmental legislation was irregular and deficient until the Australian Environment Council (AEC) and Council of Nature Conservation Ministers (CONCOM) were established in 1972 and 1974, creating a forum to assist in coordinating environmental and conservation policies between states and neighbouring countries. These councils have since been replaced by the Australian and New Zealand Environment and Conservation Council (ANZECC) in 1991 and finally the Environment Protection and Heritage Council (EPHC) in 2001.

At a national level, the Environment Protection and Biodiversity Conservation Act of 1999 is the primary environmental protection legislation for the Commonwealth of Australia. It concerns matters of national and international environmental significance regarding flora, fauna, ecological communities and cultural heritage. It also has jurisdiction over any activity conducted by the Commonwealth, or affecting it, that has significant environmental impact. The act covers eight main areas:

— National Heritage Sites
— World Heritage Sites
— RAMSAR wetlands
— Nationally endangered or threatened species and ecological communities
— Nuclear activities and actions
— The Great Barrier Reef Marine Park
— Migratory species
— Commonwealth marine areas

There are several Commonwealth protected lands due to partnerships with traditional native owners, such as Kakadu National Park, extraordinary biodiversity such as Christmas Island National Park, or managed cooperatively due to cross-state location, such as the Australian Alps National parks.

At a state level, the bulk of environmental protection issues are left to the responsibility of the state or territory. Each state in Australia has its own environmental protection legislation and corresponding agencies. Their jurisdiction is similar and covers point-source pollution, such as from industry or commercial activities, land/water use, and waste management. Most protected lands are managed by states and territories with state legislative acts creating different degrees and definitions of protected areas such as wilderness, national land and marine parks, state forests, and conservation areas. States also create regulation to limit and provide general protection from air, water, and sound pollution.

At a local level, each city or regional council has responsibility over issues not covered by state or national legislation. This includes non-point source, or diffuse pollution, such as sediment pollution from construction sites.

Australia ranks second place on the UN 2010 Human Development Index and one of the lowest debt to GDP ratios of the developed economies. This could be seen as coming at the cost of the environment, with Australia being the world leader in coal exportation and species extinctions. Some have been motivated to proclaim it is Australia's responsibility to set the example of environmental reform for the rest of the world to follow.

References

Blackman, A., Can Voluntary Environmental Regulation Work in Developing Countries? Lessons from Case Studies. *Policy Studies Journal,* 2008. 36(1): p. 119-141.

Harding, R., Ecologically sustainable development: origins, implementation and challenges. *Desalination*, 2006. 187(1-3): p. 229-239

Karamanos, P., Voluntary Environmental Agreements: Evolution and Definition of a New Environmental Policy Approach. *Journal of Environmental Planning and Management,* 2001. 44(1): p. 67-67-84.

Mitchell, R.B., International Environmental Agreements: A Survey of Their Features, Formation, and Effects. *Annual Review of Environment and Resources*, 2003. 28(1543-5938, 1543-5938): p. 429-429-461.

The California Institute of Public Affairs (CIPA) (August 2001)."An ecosystem approach to natural resource conservation in California". *CIPA Publication No. 106.* InterEnvironment Institute. Retrieved 10 July 2012.

2

History of Environmental and Biodiversity Conservation

Environmental issues have surfaced throughout human history. This comes as a surprise to many people, but the evidence is in manuscripts, publications and historical archives, and it's been there for decades.

Concern about the environment used to be found under labels like conservation, or public health, preservation of nature, smoke abatement, municipal housekeeping, occupational disease, air pollution, water pollution, home ecology or animal protection. The modern word "environmental" was developed as an umbrella to encompass a variety of related concerns. But the concerns are ancient.

The broad lack of historical perspective about environmental history has its origins in neglect and misinformation. As a result, contemporary environmental issues often emerge in the mass media without context and then disappear with little more than symbolic resolution. Political conservatives seem not to recognize the reflection of their own values in conservation movements. Political liberals lack a sense of the traditions of social reform.

Conservation Movement

The conservation movement, also known as nature conservation, is a political, environmental and a social movement that seeks to protect natural resources including animal, fungus and plant species as well as their habitat for the future.

The early conservation movement included fisheries and wildlife management, water, soil conservation and sustainable forestry. The contemporary conservation movement has broadened from the early movement's emphasis on use of sustainable yield of natural resources and preservation of wilderness areas to include preservation of biodiversity. Some say the conservation movement is part of the broader and more far-reaching environmental movement, while others argue that they differ both in ideology and practice. Chiefly in the United States, conservation is seen as differing from environmentalism in that it aims to preserve natural resources expressly for their continued sustainable use by humans. In other parts of the world conservation is used more broadly to include the setting aside of natural areas and the active protection of wildlife for their inherent value, as much as for any value they may have for humans.

Jones argues that from an economic perspective the Western nations have been no more destructive of natural resources than any other civilization. He rejects the suggestion that Christianity, by destroying animism, facilitated the ruination of nature in the West, stating he finds no evidence that any culture was or is less exploitive of the natural world than Christianity. He notes that Eastern agricultural history has numerous examples of massive deforestations, erosion, silting of rivers, and infestation with waterborne parasites. He points to large-scale animal extinction and wasteful agricultural practices by North American Indians before 1492. Jones allows that economic growth in the West did result in a higher level of resource use, but finds no evidence to support the view that such resource exploitation was a product of religion, culture, or geography.

History of Conservation Ideas

The nascent conservation movement slowly developed in the 19th century, starting first in the scientific forestry methods pioneered by Prussia and France in the 17th and 18th centuries. While continental Europe created the scientific methods later used in conservationist efforts, British India and the United States are credited with starting the conservation movement.

Foresters in India, often German, managed forests using early climate change theories that Alexander von Humboldt developed in the mid 19th century, applied fire protection, and tried to keep the "house-hold" of nature.

This was an early ecological idea, in order to preserve the growth of delicate teak trees. The same German foresters who headed the Forest Service of India, such as Dietrich Brandis and Berthold Ribbentrop, traveled back to Europe and taught at forestry schools in England. These men brought with them the legislative and scientific knowledge of conservationism in British India back to Europe, where they distributed it to men such as Gifford Pinchot, which in turn helped bring European and British Indian methods to the United States.

Sivaramakrishnan explores the boundaries between wildness and civility in Indian society, as well as connection of ideas of nature to different aspects of social life, especially labor, aesthetics, politics, commerce, and agriculture. These interconnected historical processes inform environmental history in India. At present forest history is the area of environmental history in which the most important scholarly debate is underway in India, with special interest in questions of water, air, industry, and climate change At the grass root level are organizing mass movements with the theme of Think Globally – Act locally for conservation of nature since 1993 by Vijaypal baghel, peoples are called him ecoman, greenman etc. So many events are conducting as well as Jhola Aandolan against plastic carry bags use, Global green mission, Operation water reservoir, stop global warming & climate changes, reduce pollution with dedication to save environmental and spiritual values.

Areas of Concern

Deforestation and overpopulation are issues affecting all regions of the world. The consequent destruction of wildlife habitat has prompted the creation of conservation groups in other countries, some founded by local hunters who have witnessed declining wildlife populations first hand. Also, it was highly important for the conservation movement to solve problems of living conditions in the cities and the overpopulation of such places.

The idea of incentive conservation is a modern one but its practice has clearly defended some of the sub Arctic wildernesses and the wildlife in those regions for thousands of years, especially by indigenous peoples such as the Evenk, Yakut, Sami, Inuit and Cree. The fur trade and hunting by these peoples have preserved these regions for thousands of years. Ironically, the pressure now upon them comes from non-renewable resources such as oil, sometimes to make synthetic clothing which is advocated as a humane substitute for fur. (See Raccoon Dog for case study of the conservation of an animal through fur trade.) Similarly, in the case of the beaver, hunting and fur trade were thought to bring about the animal's demise, when in fact they were an integral part of its conservation. For many years children's books stated and still do, that the decline in the beaver population was due to the fur trade. In reality however, the decline in beaver numbers was because of habitat destruction and deforestation, as well as its continued persecution as a pest (it causes flooding). In Cree lands however, where the population valued the animal for meat and fur, it continued to thrive. The Inuit defend their relationship with the seal in response to outside critics.

The Izoceño-Guaraní of Santa Cruz Department, Bolivia is a tribe of hunters who were influential in establishing the Capitania del Alto y Bajo Isoso (CABI). CABI promotes economic growth and survival of the Izoceno people while discouraging the rapid destruction of habitat within Bolivia's Gran Chaco. They are responsible for the creation of the 34,000 square kilometre Kaa-Iya del Gran Chaco National Park and Integrated Management Area (KINP). The KINP protects the most biodiverse portion of the Gran Chaco, an ecoregion shared with Argentina, Paraguay and Brazil. In 1996, the Wildlife Conservation Society joined forces with CABI to institute wildlife and hunting monitoring programs in 23 Izoceño communities. The partnership combines traditional beliefs and local knowledge with the political and administrative tools needed to effectively manage habitats. The programs rely solely on voluntary participation by local hunters who perform self-monitoring techniques and keep records of their hunts. The information obtained by the hunters participating in the program has provided CABI with important data required to make educated decisions about the use of the land. Hunters have been willing participants in this program because of pride in their traditional activities, encouragement by their communities and expectations of benefits to the area.

In order to discourage illegal South African hunting parties and ensure future local use and sustainability, indigenous hunters in Botswana began lobbying for and implementing conservation practices in the 1960s. The Fauna Preservation Society of Ngamiland (FPS) was formed in 1962 by the husband and wife team: Robert Kay and June Kay, environmentalists working in conjunction with the Batawana tribes to preserve wildlife habitat.

The FPS promotes habitat conservation and provides local education for preservation of wildlife. Conservation initiatives were met with strong opposition from the Botswana government because of the monies tied to big-game hunting. In 1963, BaTawanga Chiefs and tribal hunter/adventurers in conjunction with the FPS founded Moremi National Park and Wildlife Refuge, the first area to be set aside by tribal people rather than governmental forces. Moremi National Park is home to a variety of wildlife, including lions, giraffes, elephants, buffalo, zebra, cheetahs and antelope, and covers an area of 3,000 square kilometers. Most of the groups involved with establishing this protected land were involved with hunting and were motivated by their personal observations of declining wildlife and habitat.

History of Biodiversity Conservation

Biodiversity as a word and as a concept has literally exploded into the popular psyche and into academic literature in about twenty years. Yet the meaning of the term and the significance of related issues are often misunderstood whilst its general reference to the diversity of life and species is clear. Since its first usage biodiversity has become inextricably intertwined with nature conservation and to a degree with ideas of sustainability.

However, why and what we choose to conserve, and indeed how successful our attempts are, remain enigmatic. Ideas of conservation and biodiversity become mixed with implied issues of '*wilderness*' and of '*nature*' and '*naturalness*'. But it is increasingly recognized that much of the biodiversity which we may hope to conserve is in fact the result of a long interaction between people and nature. It is a '*cultural ecology*', the product of the environment, history and tradition.

Interactions between people and wildlife, both plants and animals, can be problematic. In many cases and across all continents, the human footprint is evidenced by catastrophic change and mass species extinctions. Today the process goes on and with human-induced climate change for example, the consequences may be even worse than previous impacts. Yet despite the damaging impacts of people on the environment and its ecology, it is clear that in pre-industrial, often subsistence societies there is an intimate long-term relationship between people and nature. Indeed, much of the ecology which we now value is present as a result of the continuity of traditional management approaches over hundreds and sometimes thousands of years. Even a cursory look at history shows that in recent centuries humankind has been rushing headlong towards urbanization and industrialization, with by the early twenty-first century over half the global population being city-dwellers. Agri-industry and agri-forestry have replaced traditional farming and woodland use, and the process of '*cultural severance*', separating people from direct dependence on nature has swept in. Yet so far, this threat to conservation has crept in with barely a murmur and most environmentalists remain unaware of the defining shift in the ecological resource.

To understand the problems we need a good knowledge of the broad issues of biodiversity and its changes through time, both natural and human-induced. However, it is also necessary to learn from history and to be able to relate human activities, economies, politics and cultures, to the changing landscape and its ecologies. From this base we can then consider the development of what we now term nature conservation and how this impacts on and even reflects the changes in the biodiversity resource. If we can gain effective insight into these processes we can then question the degree to which we have real impacts and effects. In essence, are environmentalists merely spectators who catalogue the changing kaleidoscope of nature, or do we have a real and tangible impact on the outcome? History tells us that the changes we see over centuries are reflections of deep-seated politics and economics within society; issues of supply and demand, of competition, exploitation and abandonment. This suggests that in order to alter the processes and to influence the outcomes of change in order to achieve effective nature conservation then activities must be at political and economic levels within society. If this is the case then in a democratic system it is likely that there will be serious challenges in terms of effective engagement and empowerment of people and communities in determining the nature that they desire. Indeed, these tensions are already emerging as nature conservation organizations themselves become large land-owning corporate bodies of professionals with views and agendas often imposed down on local communities rather than reflecting the aspirations of local people.

Much of the modern conservation work is of the highest excellence and has achieved restoration and remediation on a scale which twenty years ago would be inconceivable. Yet in terms of achieving genuine sustainability and a halt to human-induced biodiversity losses much of the work is fundamentally flawed. Almost all the nature conservation work is dependent on short-term grant aid and subscription-based voluntary groups and very little is embedded in long-term economic or political processes. There is little understanding by key stakeholders of the cultural nature of the resource or of the likely implications of cultural severance on biodiversity. Furthermore, there are significant issues about the lack of accountability or democracy in the determination of policies and agendas by many of the significant players. But without effective local '*ownership*' of environmental and conservation projects, any hopes of long-term sustainability must be questionable. The evidence of history and the lessons of the past inform our view of the present and the potential futures. With escalating urbanization and still growing human demands on the planetary resource the challenges will continue to raise the stakes in the environmental lottery.

Ecological diversity changes with the environment and with time, varying spatially and temporally. Over the last few millennia human impacts either directly on wildlife and vegetation, or indirectly on the environmental resources, have had major influences on ecological diversity. Each phase of human development, from hunter-gatherer, to settled agricultural communities, to industrialization and urbanization, has wrought increased human impacts on ecology and environment. Human impacts have increased dramatically in recent centuries and in the last few decades, and the effect on habitat destruction and species loss has led directly to concerns over loss of '*biodiversity*'. This was a term coined by the American ecologist E.O. Wilson in the 1980s, and which has gained widespread recognition albeit with much misunderstanding across a broad arena of debate. Whilst the impacts of early human land-use, such as forest clearance and localized agricultural intensification were undoubtedly significant and dramatic, it is in the last 500 years that the colossal scale of change has become widespread and the effects so deeply engrained.

The first big impacts of forest clearance and agricultural probably began around 5,000 years ago in areas of relatively dense population, such as around the Mediterranean and in China. There were devastating periods of drought, erosion of fertile soil, and probably major floods associated with human impacts. The deterioration of the environmental resources clearly had dramatic consequences for communities dependent upon them, but the overall effects were generally relatively localized. It is in more recent times human impact has been on a scale that can only be described as immense. This has been driven by twin processes of industrialization and urbanization, and related to these both the capability and the necessity for agricultural intensification and industrialization. It is reasonable to say that today there is no place on the planet which is not in some way affected or modified by human-related change. Human activity triggered mass extinction of fauna and flora on a scale previously unprecedented excepted in the most dramatic environmental catastrophes such as meteor impacts, mega-volcanic eruptions, and ice ages. In recent years it has also become clear that people are changing and have changed the planetary climatic condition. The debate still rages as to the precise cause and even the nature of the change but that change has occurred and a significant part is driven by human activity is clear and indisputable.

The realization of these massive impacts and of the likely or at least potential effects on the environment and what we now call '*sustainability*' has become widely accepted. Because of the increasingly visible and tangible damage to the environment and to people there has been a gradual emergence of concern and even fear about the consequences of unrestrained exploitation. Whilst this has developed gradually over the last two hundred years or so, particularly in Europe and in North America, it has emerged as a protest force and even as a political movement in the last few decades of the twentieth century and during the early years of the twenty-first. Yet concerns over ownership, use and custodianship of natural resources go back to the earliest organized communities. In animals, and especially in humans, territorial zones, often manifested as tribal areas, have been essential to survival. Much human behavior and many aspects of social etiquette are based around the establishment and protection of a resource-base. This is sustainability and survival at its most basic with communities closely and obviously dependent on their immediate natural resources. With time the human population has increased dramatically and in parallel the natural resources on which it depends have been eroded and reduced. With this scenario it has become increasingly important for individual communities to protect and ensure the survival of their resource capital, the basis of their existence. For the earliest settled communities it was increasingly vital to establish their particular niche and then avoid over-exploitation, and also to protect their resources from the depredations of competitors.

Such attempts to manage and control vital environmental assets have had mixed success over the millennia. History tells a story of over-exploitation and collapse of societies and civilizations, and of catastrophically changed environments. However, there have been successes, although many systems of resource use were not long-term sustainable. In many cases careful management regimes evolved, refined and rigorously enforced, resulted in increased production of needed material resources, from fuel-wood, to timber, to food. This was sometimes possible over long periods of time, and coppice woods for example may have evidence of active management for more than a millennium. However, whilst some systems of exploitation allowed a resource to be produced over centuries or more, they were not necessarily benign in terms of the environment. Indeed they often caused major and irreversible changes in the resource. The original ecology was often radically changed; landscapes were modified almost beyond recognition. These changes are so deeply ingrained and indeed fundamental to many ecosystems that many of our now most precious wildlife species are co-evolved with the human systems of exploitation. The flora and fauna of traditionally managed grasslands and commons, from hay meadows to pastures are examples of these rich cultural systems. But when tradition ceases and economically-driven management ends, the landscape and its associated biodiversity deteriorate. This theme will be re-visited later.

Yet many exploitation systems have proved fatally flawed and as a result societies and even major civilizations collapsed. Underneath many if not all of the world's most extensive deserts for example, lie the ruins of a failed civilization; they over-exploited their resources and perished as a consequence. Forest clearance around the Mediterranean led to serious shortage of essential fuelwood and timber and declines in early civilizations. The clearance of upland forest and the cultivation of lands in Britain can be evidence by down-washed sediments in lowland floodplains and areas such s the East Anglian Wash. Over-exploited in Bronze Age and Iron Age times, the skeletal soils simply eroded and leached away to leave a nutrient-depleted ecology and a permanently changed landscape. Over-use of Spanish forests in medieval times meant the loss of timber needed for ship-building and a long-term decline in both economic and political power. In many if not most cases the consequences of such catastrophes have proved long-term and irrecoverable.

Humans exploiting, foraging in and managing environmental resources have influenced ecosystems and biodiversity for millennia. Competition for resources has also been pivotal in the balance between different communities and civilizations. Indeed the impetus of gaining specific resources was often the key driver for conflicts and conquest. This was the case from the Roman Empire to European imperialism and expansion in the post-Renaissance period and beyond.

Natural Variations in Ecology and Biodiversity

Not all fluctuations in ecological diversity are attributable to human activities. Observations of natural ecosystems show how ecology varies spatially with environmental variables like soil, geology, water, topography and aspect, and climate. Variation also runs through time as distribution and ecological successions generate change, and over longer time-periods through processes of evolution and extinction.

This latter point leads to one of the key misunderstandings in terms of extinction and biodiversity. A common and popular fallacy is that since extinction is '*a natural process*' and '*species have always come and gone*', then human-induced destruction of wildlife is itself '*a natural process*'. Whilst species loss can be a natural process the key issues are the *rates* of extinction and the *scale* of human-induced losses, and especially of recent impacts. The recent changes amount to a mega-extinction and although it is true that mass extinctions have occurred previously, they have generally been associated with globally catastrophic events such as meteor impacts, mega-volcanic eruptions, and long periods of ice age. The present rates of human-induced extinction are comparable with those of earlier catastrophes. This should be a salutatory fact and not a reason for complacency.

Imprints of Environmental Change on Biodiversity

One of the most obvious influences on biodiversity, and of great concern today, is climate change. The effects of long-term climatic fluctuations are visible in ecosystems around the globe. Furthermore, the impacts of human-related changes in for example, forest cover and the spread of deserts through over-grazing, are frequently inter-locked with natural fluctuations. These changes are then reflected in biodiversity at every level from local and regional to global. One of the most pertinent examples of climate change and changed biodiversity, especially for those in the Northern Hemisphere, is the so-called '*Little Ice Age*' from the 1400s to the 1800s. It is worth considering this episode in a little more detail.

An Example of the 'Little Ice Age'

The causes of the major deterioration in the climate of the Northern Hemisphere are still disputed. However, whatever the reasons, the result was a period of several centuries of incredibly harsh cold weather. It is difficult to be precise about the scale and the nature of the impacts on biodiversity because there was almost no scientific recording and therefore very little evidence of the detailed ecology of the time. However, it is clear that wildlife was squeezed out of northern regions and into more southerly, warmer areas. There must have been significant losses of climate-sensitive species particularly in the northern areas. We get some idea of the likely changes through the retreat of the growing areas for particular crops and fruits such as the Grape Vine (*Vitis vinifera*), and also the widespread failures of harvests across Europe during this period. Some indication of biogeographic trends can be seen in the southward spread of northern species such as for example the Hooded Crow (*Corvus corone cornix*) and its more recent retreat back north. Other sources of information are locked away in the paleo-ecological archives of peat bog profiles in which the loss of warmth-requiring beetles as the icy grip of the Little Ice Age squeezed them out. Similarly there may be evidence from long-lived trees such as Small-leaved Lime (*Tilia cordata*), of previous climatic conditions and of extreme events. The evidence is relatively thin and it is also hard to separate the climatic trends and induced changes from those caused by other aspects of landscape change including human agricultural practices.

The suggestion would be that northern and boreal species of pants and of wildlife would have spread south during this time and that southern thermophilic species would have moved south too, and many may have been lost. Certainly if we consider a national biodiversity resource such as that of the British Isles, then there would have been undoubted gains and losses on a large scale. The extent to which the European fauna and flora would have gained or lost is very difficult to predict.

Effects of Some Other Major Environmental Impacts

Global and regional biodiversity has been grossly affected by the impacts of major volcanic eruptions, by both direct regional or local effects of volcanic ash, gas clouds and lava, and by the discharge of particulates to the high atmosphere which caused intense periods of cold and dark weather. Some of these events are now known to have caused huge suffering amongst human populations and the abandonment of for example, more marginal settlements in upland areas. They must have impacts on biodiversity, but aside from the obvious removal of fauna and flora in the immediate vicinity of a volcanic eruption, there is little known about the detailed effects on biodiversity. Edward Wilson discusses the likely implications of the known major volcanic events on biodiversity. He also considers the impacts of major meteor impacts and of course these are now recognized as being responsible for mega-extinctions of fauna in prehistoric times. One of the major events of this kind occurred around 66 million years ago and was responsible for the extinction of the dinosaurs. Whether this was a single event or perhaps a series of smaller meteor impacts and also volcanic eruptions is still, debated. Extinction did not happen all at once but was spread out over a few million years.

Fluctuations in sea level have also occurred since the first oceans were established on the Earth. It is considered likely that major and cataclysmic sea level changes and consequent flooding occurred in the early periods of the emergence of major civilizations and are '*recorded*' in the flood mythologies and folklore around the world. The most widely-recognized are the accounts of the biblical flood which may relate to the post-glacial filling of the Mediterranean and then the Black Sea as the Atlantic rose and water poured through to inundate the previously dry basins. Again we can only speculate as to the impacts on ecology and biodiversity of events of such magnitude.

However, flooding and sea level related changes have occurred throughout history and prehistory. The East Coast of England for example has suffered major periods of inundation, of coastal zone erosion, and also of massive deposition and accretion. All these have affected the ecology and biodiversity of the region. Even the process and pattern of change have influenced our contemporary fauna and flora, as for example, Britain was cut off from continental Europe by rising water around 7,000 years ago. The precise timing of sea level rise was hugely influential set against the process of northward re-colonization of animals and plants in the post-glacial period as climate generally warmed. The Pine Marten (*Martes martes*) was a mammal, a mustelid, which made the move in time to get across before the English Channel closed and is a part of the British fauna. The Beech Marten (*Martes foina*), its southern cousin failed. The tiny Harvest Mouse (*Mycromys minutus*), our smallest mammal is believed to have crossed what is now the North Sea in the reed-like vegetation which lined the path of a great river running across the shallow flatland between Europe and Britain.

Areas of massive coastal flooding would be changed totally and some zones became wetter and other areas gained new ground. Such regions would have had extensive marshes and flats, and also vast areas of dynamic sand-dune systems with all their various successional stages. It is likely that many of the species adapted to such dynamic landscapes were the so-called ruderals, often today's '*weeds*'. The present-day natural habitat of such species has often diminished, though as noted by Oliver Gilbert (1989, 1992) they or their cultivated descendants are often important in urban commons and other disturbed environments. Indeed, the successional changes and sequences of a regenerating urban or post-industrial disturbance site were compared by Gilbert to those of the post-glacial re-colonization of Northern Europe; albeit telescoped in time sequence from many centuries down to a few decades.

Ecological Footprint of the Human Race

It is generally accepted that the major process of re-colonization of the Northern Hemisphere reached its zenith around 5,000-6,000 years ago, with many of today's main vegetation zones and ecosystems in place. However, since that time, with both successional change and in response to long-term and short-term climatic fluxes, the situation has remained dynamic. Species' distributions across the region and hence the biodiversity of any one area have changed dramatically over the centuries. In trying to assign species and biodiversity to specific areas or regions, it is sometimes hard to take the inherent dynamic nature of the resource into account. Such changes occur today, with for example the spread of collared dove across Europe from the East during the latter half of the twentieth century, and the more recent colonization, or re-colonization, of Britain by Little Egret (*Egretta garzetta*) and Cattle Egret (*Bubulcus ibis*). Despite such changes it is clear that the basic structures and diversity of ecology across Northern and Western Europe were in place by a period perhaps several thousand years ago. Since that time, when human activity was relatively limited and had effects through the use of fire, and also through indirect impacts via influences on large grazing herbivore behavior, the human footprint has become massive. Early peoples certainly influenced the landscape and as in North America, the selective hunting of large game would definitely influence savannah and forest growth, species and patterns. Yet potentially important though some of these influences might have been, the phases of human social development that followed make the earlier impacts seem miniscule. There is still a debate about the nature of this primeval landscape across North Western Europe, and the discussions centered on the seminal work of Frans Vera, have helped to clarify how the early landscape might have looked and perhaps aid the understanding of how their biodiversity relates to that which we see today.

Our vision now is not one of wall-to-wall forest, but of more open plains or savannah with a rich diversity of other landscape and ecological components too. Much of this variation would be related to basic environmental factors like climate and water-logging, but Vera's key factor is the importance too of large grazing herbivores. Not everyone agrees with Vera's hypothesis, but the argument is compelling and a refined version of his original vision has a considerable body of supporting evidence.

The landscape had vast expanses of wetland, marsh, fen and peat bog with extensive coastal wetlands too. Dense woodland and thickets would grow up in the protection of rings of prickly Blackthorn and Bramble and here would be the plants and animals that characterize our so-called '*ancient woodlands*' today; the Bluebell (*Hyacinthoides non-scripta*), Dog's Mercury (*Mercurialis perennis*), Wood Anemone (*Anemone nemorosa*), and Yellow Archangel (*Galeobdolon luteum*) for example.

Outside the prickly halos of the thorns was a wide open savannah with heath, grassland and giant old trees such as Pedunculate Oak (*Quercus robur*), to ages of a thousand years or more before collapsing into oblivion.

A further powerful driver of succession and change in this landscape would be pre-human fire caused by lightning strikes which would take out the great trees in the open plain. Further north and into the upland zones there would be a change to landscape dominated by great and similarly ancient Scots Pines (*Pinus sylvestris*), and again a strong influence of natural, lightning-related fires.

Mountain zones would have large areas of disturbance through natural landslip and erosion areas, and all the great rivers would include sometimes vast floodlands and meandering patterns of erosion and deposition in an ever-changing yet stable landscape. Much of this environment would have limited amounts of available nutrients, especially nitrogen and phosphorous, and this too had a great influence on associated biodiversity.

Localized areas such as mountain downwash zones and alluvial fans would have higher nutrient levels, and the whole landscape would have abundant micro-disturbance through natural process, but only limited macro-disturbance.

Erosion and deposition areas, animal-related disturbance, fire, and successional and life-cycle related changes such as the collapse of ancient trees provided a template for a richly diverse fauna and flora with regionally district ecologies related to broad climatic influences and localized geological and topographic factors.

This was the landscape, the ecology and the biodiversity upon which the footprint of human activity would be stamped with increasing effect over the following five to six thousand years.

References

Agnoletti, M. (Ed.) *The Conservation of Cultural Landscapes.* CAB International, Wallingford, Oxon, UK.

Beard, C. *Wildlife Conservation and the Roots of Environmentalism. The Facts and Figures.*

Freedman, B. *Environmental Ecology – The Effects of Pollution, Disturbance and Other Stresses.* Second Edition, Academic Press, San Diego.

Gaston, K.J. (Ed.) *Biodiversity: A Biology of Numbers and Difference.* Blackwell Science, Oxford.

Jeffries, M.J. *Biodiversity and Conservation.* Routledge, London.

Riley, D. & Young, A. *World Vegetation.* Cambridge University Press, Cambridge.

WCED *Our Common Future.* The World Commission on Environment and Development. Oxford University Press, Oxford, UK.

3

Community-Based Environmental Protection

Over the last twenty-five years, federal and state anti-pollution laws have achieved many notable environmental successes. Local communities often play a prominent role in addressing many of the most pressing environmental concerns. Central to these concerns is the need for clean and vital ecosystems. Healthy ecosystems support human health, plants, and animals. They also provide recreational opportunities and support local economies dependent upon fish, game, forests, and other resources. Full protection of our nation's ecosystems requires communities and individuals to conserve or restore habitats and solve other environmental problems not specifically addressed by traditional regulatory approaches.

Over many years, a number of communities in the country have initiated their own successful community-based environmental efforts. Indeed, the first anti-pollution laws around the turn of the century were local ones. This publication draws on the experiences of many different communities to provide examples of community-based environmental programs and key approaches, information, and other tools that communities are using.

Community-based Environmental Protection

Community-based environmental protection is action that local individuals and groups take to address their own environmental concerns. Ecosystem protection carries such activity beyond localized environmental issues, such as pollution from a particular factory or lead poisoning from paint in older housing, to consider the ecological health of the total local environment. This environment often extends beyond municipal borders.

People who work, live, and have businesses in the community have a common interest in protecting their shared environment and quality of life. The defining element of community-based ecosystem protection is that these people work together to develop plans and goals. Ecosystem protection plans developed in this way can be very effective because:

- They take into account local social, economic, and environmental conditions as well as community values.
- They create a sense of local ownership of issues and solutions and encourage long-term community support and accountability.
  - Many local economies depend on outdoor recreation and tourism.
  - Many communities rely on resources extracted from the environment. Examples include timber, minerals, building materials, and seafood.
  - Ecosystem quality affects the value of property and may influence local finances.

The importance of ecosystem quality to sustainability is illustrated by considering how community members' economic lives are made possible by healthy ecosystems. While modern technology has provided many substitutes and supplements to goods produced by healthy ecosystems, such as farm-grown fish rather than wild fish, the economy is still dependent upon the environment for basic raw materials like water, wood, and minerals. For the economy to grow, communities must protect the underlying natural systems on which they are built. For example, if a community harvests timber in a given region faster than the timber can grow back, or if a local shellfish bed is over harvested, the industries that depend on these resources will fail. Careful management of ecosystems such as forests and estuaries can help avoid this outcome and provide a continuing supply of products, services, and jobs for the next generation.

This publication is intended to help you understand how healthy ecosystems benefit your community and how recreational, economic, and other activities affect the quality of your ecosystems. It will show you how other communities have assessed the interrelationships between their community goals, such as residential development, and ecosystem quality.

As well, it will show you how those assessments helped communities decide how to focus their efforts and resources more successfully.

Goal-Setting and Developing an Organization

Getting Everyone Involved

Community ecosystem protection initiatives often begin at the grassroots level, when friends and neighbors share a common interest in protecting or restoring the local environment. These initiatives may be spurred by noticeable air or water pollution, a development that causes ecosystem damage, some obvious ecological effect such as a fish kill, the gradual loss of desired species such as songbirds, or some other symptom of an underlying ecological problem. Alternatively, a community might come together to protect local ecosystems before they become threatened.

A concerned citizen, local official, or other project initiator may have some idea of desired outcomes, or may have identified ecosystems or ecosystem components to improve or protect. Project initiators in other communities have found it useful to reach out early to other *stakeholders* — meaning, literally, people who have a stake (or at least an interest) in what the initiator is thinking about — to begin an exchange of ideas about the desired outcomes or conditions that sparked their interest. Identifying possible stakeholders and sharing information stimulates their thoughts and desire to participate Ultimately, stakeholders develop partnerships by coming to agreement on issues, vision, and information, leading to the development of a set of community goals and actions.

Who are possible stakeholders? They include anyone in the community who takes a natural interest in environmental protection. Groups that might be affected by changes in commercial activity resulting from ecosystem protection strategies are also potential stakeholders. Examples may include businesses or labor unions. Local elected officials and community leaders can help identify potential stakeholders, in addition to participating themselves.

Potential stakeholders might include the following organizations and individuals.

a. Members of existing organizations that use or are concerned with the environment or land-use issues

b. Private landowners whose property includes habitat areas that the community wants to protect, including farmers, ranchers, timber companies, and private residents

c. Businesses whose livelihoods depend on local environmental resources, directly or indirectly

d. Local chapters of relevant national professional organizations

e. Offices of state, local, tribal, and federal governments

f. Faculty at local schools and universities, especially those in environmental studies, biology, ecology, geology, and other natural sciences as well as economics, urban planning, public policy, and other social sciences

g. Senior citizens' organizations such as local councils on the aging or the Environmental Alliance for Senior Involvement (EASI).

Many diverse ethnic, religious, or other groups might be interested in sharing their points of view and participating. In some cases, communities must actively seek the involvement of key groups. Stakeholders may exist outside the immediate geographic area.

Often, a community's ecosystem protection effort will interest people who live in distant places. For example, a land conservation effort in a rural resort area may cap ture the interest of city-dwellers who spend summers there. Similarly, a river restoration effort may affect many downstream communities. The economic interests of people in other areas also may greatly affect communities' efforts. The following stories about *Flagstaffs Open Spaces and Greenways Committee* and the *Anacostia Watershed Restoration Committee* illustrate efforts to include new members.

Engaging people from all key stakeholder groups as soon as possible produces many benefits. People are much more likely to work together successfully if they are involved from the beginning rather than after decisions are made. For example, developers may be more willing to discuss alternative development schemes if they are invited to help plan ecosystem protection strategies. Many community members gain a sense of well-being from volunteering their time to create a better community; involvement in the effort can be a source of personal enrichment.

Most communities have found that communication is vital in getting stakeholder involvement. For exam-pie, visiting some of the groups noted above at one of their meetings and speaking for five minutes might successfully draw stakeholder participation.

Identifying Goals and Defining an Approach

Other communities have found that when they first embark on an ecosystem protection project, they do not have a clear idea of goals, other than a general concern about protecting local ecosystems. Various methods of goal-setting, such *as visioning*—forming a concept of what the ideal state of the community's ecosystems should be—can help a community develop goals.

Setting Versatile Goals

When thinking about goals, many communities have considered not only ecological protection, but also the ways in which the environment interacts with quality of life and the local economy. These three endpoints can guide goal setting. For example, your primary ecological protection goal might be protecting streamside or woodland habitat. An associated "sustainable economy" goal might be working with landowners to preserve their woodlands by carefully planning and selecting the timber harvest to protect tree age, size, and species diversity, and replanting species native to the area. Improving the quality of life might combine protecting wildlife habitat with construction of nature trails to provide hiking and walking benefits. Goals to improve ecosystems can include both present and future generations — that is, the ecological legacy the community wants to leave its children and grandchildren.

Using Indicators to Measure Progress Toward Goals

Tying goals to *indicators,* or specific measures of how well the community is achieving its goals, is a concrete way to determine progress. For example, measuring community progress in protecting aquatic species might involve counting the number of wading birds in the area. Specifically, the community could seek to double the wading bird population by the year 2000. Measuring the economic health of the community might involve tracking employment in eco-tourism businesses with *a* goal such as 50 percent growth in local eco-tourist business *by* the year 2000.

Adapting Goals

As more and more stakeholders join the effort, the community may need to go through the goal development process more than once. The period after the assessment, planning, and execution of a particular ecosystem protection project provides an opportunity to reevaluate whether the community is meeting goals, using indicators, and whether these goals indeed represent the priorities of the community. If not, then the strategies chosen may not be effective and the underlying goals may not be relevant and realistic. Ultimately, the community may want to undertake another goals development session.

Making Sure the Goal-Setting Process Involves the Whole Community

Unless key community members participate in setting goals, the goals produced will not legitimately reflect the wishes of the community as a whole. Goal-setting works best when participants are an inclusive group.

Defining Geographic Boundaries

As part of developing goals, outlining problems, and developing solutions, communities have found that they need to consider the boundaries of the ecosystem that they wish to protect. Determining these boundaries is not always straightforward. In particular, it involves understanding the complex interactions between people and their environment.

Starting Small

The boundary-drawing exercise is complicated by the fact that most ecosystems are not wholly self-contained. A lake, for example, may be a component of a larger nat-ural system of rivers and streams within a watershed. Therefore, the lake may be affected by runoff, pollutant spills, flooding, and other problems affecting related water bodies. Some ecosystems are so complicated that it may be difficult to address the entire system. River deltas, with their networks of fresh and saltwater marshland and rivers and lakes, are an example of such a system. Furthermore, ecosystems such as forests or lakes often cross town, county, and even state boundaries.

These considerations may discourage communities from going forward with ecosystem protection plans. The community may feel that its ecosystems are so interconnected with the larger environment that whatever small steps it takes locally will be overwhelmed by events occurring in related ecosystems in other towns or states. Alternatively, the community may keep increasing the area of interest to incorporate as many ecosystem features as possible, then realize that it will need to reach out to other communities for their cooperation.

Some communities have found it useful to start small. Considering ecosystems in the context of the larger environment of which they are part doesn't require tackling the entire system at once.

Sometimes, however, retaining a small geographic scope may not be feasible Expanding the scope can help include the following:

- *A Critical Locale* - For example, a project may be more effective if it covers an important tributary to a river or a woodland that contains a crucial nesting site for birds.
- *A Critical Stakeholder* - For example, a large landowner may be able to make a significant contribution to the health of local ecosystems through land management techniques.
- *Special Skills or Resources* - The community may want to expand boundaries, for example, to make the project relevant to a nearby university or to include endangered species habitat that will capture the interest of federal agencies.
- *Special Constituencies* - The community may want to expand boundaries in an explicit effort to include, for example, groups who historically have been overburdened by environmental degradation or have been systematically left out of other community decisions.

The Big *Sandy Lake Association,* discussed on the next page, shows how one community expanded its ecosystem protection effort.

A community can use the boundary-drawing exercise to help in thinking about other towns, counties, or even states with which to cooperate. If a community is considering making a river swimmable, for example, the effort will be affected by what goes on upstream. For this reason, communities often work closely with the watershed association and state entities, and may also involve other towns.

Deciding What to Include

Communities often start with the most obvious ecosystem unit and enlarge the area of interest by considering related ecosystems. If a community is focusing on a small pond, for example, it may also consider including wetlands, marsh, or wooded areas around the pond. Outlining the area on maps clearly shows topography (for example, elevation, water bodies, and other features) as well as political features (for example, roads and state, county, and city boundary lines). The *California Natural Communities and*

Conservation Planning Program illustrates one group's efforts to define the boundaries of a local ecosystem.

In drawing boundaries, many projects have considered whether to include a buffer zone around the ecosystem. Such a zone absorbs the effects of human activity around the core of the ecosystem, preventing damage to the sys tern itself. For example, for a seacoast, a zone of non-marsh, non-sandy terrain between the water's edge and development can prevent erosion and protect delicate tidal ecosystems. A project conducted by a citizens' group in *The Berkshires of New England* is an example of the successful implementation of a buffer zone around a river.

Choosing the Best Organizational Structure

Many communities function well as an ad hoc collection of members with shared responsibilities, each participating in tasks or responsible for implementing a part of the plan. Other communities decide to implement a formal organizational structure. Elements of such a structure might include:

Steering Committee—The steering committee assumes the day-to-day responsibilities of the organization, delegates tasks, and may be responsible for outreach to other organizations. *The Blackjoot Challenge* and *Tensas* River *Basin* stories, on the following pages, provide examples of groups that exchanged ad-hoc structures for steering committees.

Legal Incorporation—Becoming a legal non-profit entity has advantages for some organizations. They include tax exempt status, access to certain grants, and protection from personal liability for group members. Incorporation involves establishing a board of directors and organization bylaws; an attorney generally handles the incorporation process.

To summarize, the following story about *the Neponset River Watershed Association* illustrates many of the elements involved in initiating a community ecosystem protection plan.

Assessing the Conditions of Local Ecosystems

A doctor uses blood pressure, body temperature, and other data to monitor a person's health. In the same way, communities can assess and monitor the health of their ecosystems by collecting and analyzing various kinds of "indicator" data. A wide variety of indicators might be used. The kinds of indicators that are important depend on the characteristics of a community and its priorities.

It r describes what indicators are and explains their relevance to community ecosystem protection. Most communities have considered three types of baseline assessments:

- Assessing the health of local ecosystems and identifying factors ("stressors") affecting their quality
- Assessing links between local ecosystems and the local economy
- Assessing links between local ecosystems and the community's quality of life.

Using Indicators

What Are Indicators?

Indicators are measures that help you assess the health of local ecosystems, understand what factors are affecting ecosystem quality, and assess the effects of ecosystem quality on life in your community. Indicators provide insights into the condition, qualities, interrelationships, or problems of a complex system, such as an ecosystem or the local economy. Over time, tracking an indicator helps to measure progress toward a desired goal.

Most indicators are *quantitative*: they are numeric data based on actual measurement of the factor being monitored. *Qualitative* indicators generally attempt to describe a factor of interest, rather than measure it with precision. An example of a qualitative indicator would be a description of a local lakefront as "moderately polluted" or "very polluted". Qualitative indicators are valuable because they can describe situations that cannot be measured with a single data series.

The accuracy and reliability of qualitative indicators depends on the knowledge and biases of the people providing information. Qualitative indicators can be difficult to interpret and may mean different things to different people. In contrast, quantitative indicators are less subject to conflicting interpretations.

Three general kinds of indicators are discussed:

- *Indicators That Characterize Environmental Health*—For example, the number and variety of grass species is an indicator of a prairie's ecological health. Because people are part of the ecosystem too, indicators of their health and safety are also useful.
- *Indicators That Reflect the State of the Local Economy*—These indicators track the economy as it is affected by the quality of

ecosystem resources and services; for example, the number of people employed in commercial fishing or in industries that depend on a clean water supply (such as breweries and food processing).

- *Indicators That Reflect the Community's Quality of Life*—These indicators track quality of life as it depends on ecosystem quality, such as the number of visitors to a public beach or levels of traffic congestion and vehicle miles traveled.

Why Use Indicators?

Using indicators is a shorthand method for obtaining representative information on the overall system. Each single indicator reflects only a part of the complex system. When indicators that measure key aspects of the system are looked at as a set, however, they reveal trends and interrelationships that might not otherwise be apparent. For example, declines in bird populations on a lake shoreline in combination with data showing increased boating activity on the lake might suggest a cause for the reduced bird counts. Note, however, that comparing two data series does not *prove* a cause-and-effect relationship between them.

Indicators provide a relatively objective basis for discussion, planning, setting goals, and measuring progress. They help avoid the misdirected effort that might result from simply reacting to the most obvious trends or relying on a few people's untested opinions about what actions should be taken.

Choosing Indicators

Some indicators address the community as a whole system — ecological, social, and economic. For example, while the amount of fish taken by a commercial or sports fishery may be of interest, this indicator may say little about the health of the aquatic ecosystem. Some more targeted indicators would include information on the presence of tumors in fish, the numbers of fish within age classes in the population, and the availability of their food resources—as overall indicators of the fishery's health. These indicators help to measure the sustainability of the local fishery, thus measuring both economic security and ecosystem stability and quality. For the *Santa Monica Bay Restoration Program*, citizens developed a list of potential indicators when evaluating what types of data were needed to create a comprehensive monitoring system for the bay. Their list illustrates the many characteristics different people considered when they thought about the health of one specific ecosystem.

Often, indicators need to be understandable and useful to a range of audiences, including the general public, public officials, and scientists. In communicating with the general public, less technical measures are often preferable. On the other hand, communications with public officials or scientists, perhaps when seeking future funding, often benefit from the use of more technical language. The difference may simply be in the wording of the indicator. For example, scientists wanting to protect water quality might find it useful to know specific bacteria levels in water. The general public, however, may find more useful the percentage and location of rivers and streams considered unsafe for swimming because of high bacteria levels.

One way to select indicators is by brainstorming with all interested parties to identify an ideal set, keeping in mind what is being measured. Ways to narrow down a list of possible indicators include looking at data sources, investigating sources of help, and deciding what information is most useful. Generally, monitoring a few key indicators well provides more useful information than monitoring a wide variety of indicators poorly. For example, a community may want to measure the recovery of an aquatic ecosystem by sampling the number of different types of benthic organisms (bottom-dwelling species such as worms and shellfish). Such a field survey may require a large budget, however. Instead, the community could track fish abundance or water quality as a proxy for ecosystem recovery. Also, data may already be available for developing certain indicators but not others; a less than perfect indicator supported by available data may be more practical than an ideal one that requires extensive data gathering.

In summary, good indicators will reflect stakeholder concerns, be readily understandable to their audience, be responsive to change in the ecosystem or community, and be appropriate for highlighting emerging ecosystem problems before they become irreversible. The *Sustainable Seattle Program* below provides a good example of effective community involvement in indicator development and selection.

Assessing Conditions and Trends in Local Ecosystems

Your community may have begun its ecosystem protection effort in response to some actual or potential threat. Perhaps you have observed a decline in the number or diversity of birds that inhabit a local forest, or have grown concerned that open space in your town is disappearing due to sprawling residential development.

Indicators of ecosystem health serve two purposes. First, they define the problems you are trying to address. Second, they track progress over time from a starting point or "baseline". Collecting consistent data over time and comparing them to the baseline data enables an evaluation of whether the actions taken to protect ecosystems are actually working.

Specific Indicators of Ecosystem Health

Community characteristics (such as urban vs. rural, coastal vs. inland) and environmental problems affect the choice of ecosystem health measures. For example, a community may be interested in determining the degree of biodiversity in local ecosystems. A specific indicator would be the number of bird species found in an annual bird count. This number could be compared to counts from previous years to evaluate the trend in biodiversity.

To start defining ecosystem assessment objectives, it might be useful to consider the physical, biological, or chemical changes the system has undergone. Relevant questions may include:

- *Physical Changes to the Ecosystem*—How has the structure of the system changed? Has the number and/or kind of habitat types in the area changed? Has the size of a forest area declined? Has the amount of water flowing through a river changed? Are wetland areas shrinking? How much low-density development of "greenfields" (such as farmland, forests, meadows, open space) is occurring? How much development is occurring on parcels of land not adjacent to existing urban areas? How much habitat remains of the original area? Is the remaining habitat fragmented?
- *Presence of Harmful Chemicals*—Are toxic chemicals, excessive nutrients, or other pollutants present in the soil, air, or water, or in the plants and animals living in the ecosystem?
- *Biological Damage to the Ecosystem*—Physical and chemical changes to the ecosystem are likely to produce changes in the plants and animals that are

Developing an Historical Perspective

Analyzing the historical condition of an ecosystem helps assess its current health. Identifying man-made and naturally occurring forces that have affected the ecosystem also can help set sensible project goals, providing reference points for measuring progress.

For example, publicizing that a severely degraded local ecosystem once supported a large number and variety of birds and other wildlife can act as an incentive for local officials, business people, environmentalists, and the general public to restore it. Moreover, by understanding the system's history and the ways in which it was exposed to different stressors, communities can help evaluate the relative impact posed by these stres-sors. For example, a decline in wildlife populations may be caused more by a dam placed on a major river than by pollution associated with industrial activity. Understanding how events led to the current condition helps communities weigh alternative actions for restoring and protecting the ecosystem.

Other communities have found the following information sources useful in seeking out historical information on local ecosystems:

- *Surveys of the Community and Anecdotal Information*—Other communities have found a survey of the community to be very valuable. A survey not only will gather the public's ideas on how the quality of the ecosystem has declined, but will help publicize your project. In addition, informal conversations with older residents can be an excellent source of anecdotal historical information.
- *Local Historical Information*—Depending on the size and location of the community, a number of institutions may be able to provide information on the community's history and the evolution of the ecosystem. Possible examples include the local library, town or regional historical societies, colleges and universities, and local non-profit organizations such as land trusts. If useful local histories do not already exist, a local high school or college student may be able to develop a project on ecosystem history. Deeds and maps from the county registry of deeds or municipal offices also may provide historical information.

Gathering Technical Data

Once you have a good idea of how the ecosystem arrived at its current state, you can assemble a more detailed and rigorous evaluation of current conditions. This assessment will allow your community to identify those components of local ecosystems that currently are degraded or at risk and evaluate the extent of the problem. This will provide a baseline "snapshot" of ecosystem health against which protection measures can be evaluated.

Local, state, and national agencies collect an enormous amount of environmental data. The information ranges from observations about the general appearance of an ecosystem to detailed analyses of emissions of toxic compounds from industrial sources.

Role of Government Agencies

State or federal environmental agencies can help locate data to assess ecosystem health. While the workings of the agencies devoted to environmental protection may be unfa- miliar, a few strategic phone calls may take you far. State environmental agencies often have published reports on statewide environmental conditions. These "environmental indicators" or "state of the environment" reports provide a useful overview of trends in the state and may point to other data sources. The state agency is also likely to be familiar with the local area and should be able to offer guidance on other regulatory agencies (such as federal organizations or other state and local agencies) that can help further.

Role of Non-Governmental Organizations

In addition to government agencies, a large number of non-government organizations provide information to assess ecosystem quality. For example, conservation groups such as The Nature Conservancy and the National Audubon Society and state Audubon societies may be useful in assessing the quantity and quality of local wildlife habitat. Grant programs administered through universities may also be an important source of information.

Role of Local Resources

A variety of local resources also may prove helpful in searching for ecosystem assessment information. For example, local public health agencies may have information on swimming advisories associated with local beaches, and may even keep more detailed water quality data. Data on soil and/or ground-water quality may be available through real estate transaction records that require environmental inspections prior to property sales. The results of such inspections may be available from the local registrar of deeds or real estate board.

Linking Stressors with Impacts

After characterizing the current state of the system and noting areas of concern (such as wetland loss, decline in plant or wildlife species), it is possible to identify the sources of those problems. For example, natural fish populations may be in decline, but why? Is it due to chemicals from industrial dischargers, over-fishing, or the damming or channelization of the river? Stress on an ecosystem can come from a wide range of sources, including industrial and municipal sources discharging toxic chemicals, agriculture and livestock feedlots, petroleum and chemical storage tanks, mining (for instance, acid mine drainage), recreational activities (such as stream bank erosion caused by boats and jet skis), water withdrawal by industry and utilities, septic tanks and other development impacts, and waste management.

Links Between Ecosystems and the Local Economy

Local Economies Depend on Ecosystems

The Federal Interagency Ecosystem Management Task Force has said that sustaining the health, productivity, and biological diversity of ecosystems, "is essential to maintain the air we breathe, the water we drink, the food we eat, and to sustain natural resources for future populations." As this quote indicates, our lives are greatly influenced by the healthy functioning of ecosystems.

Ecosystem Components May Have Direct Commercial Value

Much of our country's wealth is the result of an abundant supply of natural resources and the ecosystems that sustain them. The link between ecosystems and the economy is clearest in communities that extract renewable resources from the environment. For example, the economy of many communities in southeast Louisiana is highly dependent upon shellfish beds and the larger system of wetlands that surrounds and protects them.

The issue of "jobs versus the environment" often arises when discussing ecosystem protection. Controversies such as limiting forestry to protect endangered species have led many people to believe that black-and-white choices must be made between resource extraction or land development and ecosystem protection. In fact, ecosystem protection is often pursued when a community is looking for ways to manage its resources and sustain local industries. While this may sometimes result in short-term reductions in economic activity (such as limitations on the commercial fishing catch), the resource may be maintained for the long run, making the local economy more

Ecosystems and Quality of Life

Healthy Ecosystems Make Life More Fulfilling

Healthy ecosystems make our communities more rewarding places to live in ways that are unrelated to economic conditions. Many of these benefits are subtle, and different communities may emphasize different aspects of well-functioning ecosystems depending on their values. They include:

- *Natural Beauty*—Natural areas provide a sense of well-being for the community. In particular, protecting habitats in urban areas gives citizens an opportunity to "leave the city behind", view local animal and plant life without having to travel, and gain a sense of inspiration and renewal.
- *Protection of Human Health and Safety*—Clean air and water and healthy ecosystems ensure that the community is free from health problems associated with pollution. In addition, community members have the peace of mind of knowing that they are safe from these threats.
- *Sense of Community*—A healthy natural environment enhances feelings of civic pride and may instill a stronger sense of kinship among residents.

Natural spaces can be used for community gatherings such as annual festivals, picnics, graduation ceremonies, and community gardens. The collective action necessary to protect ecosystems can itself be a bonding force and source of pride to residents who share a common goal and work together to make their community a better place to live.

- *Spiritual Value*—Many people feel spiritually connected with the ecosystems around them. The beauty of nature gives them an opportunity to contemplate their relationship with the world. Many religious denominations have organizations that promote environmental stewardship because of the belief that humans have a special responsibility to protect and pass on a healthy world. Many of these groups see the principle of sustaining the health and benefits of the natural world for future generations as a moral obligation. Native American cultures have often been identified as placing a particular spiritual and religious significance on nature and the harmony in natural systems. Many people in the United States share these values.

- *Educational Value*—Rivers, wetlands, forests, and other elements of community ecosystems all provide learning opportunities. Certain areas may be designated public learning centers by the town or by conservation groups. Local colleges and universities also may use ecosystems for teaching and scientific research. For schoolchildren, healthy ecosystems may provide a sense of wonder and encourage further learning. Many parks and forests make learning easy by posting illustrations and names of local plants and animals at park entrances or on placards along trails.
- *Recreational Opportunities*—Healthy ecosystems support wildlife and other natural resources that are often central to outdoor recreational activity. For example, wetlands may provide breeding and spawning areas for fish sought by recreational anglers and may support bird species popular with birdwatchers or hunters. In addition to the commercial value of recreational resources, the intrinsic value of the recreational experience is an important part of life in many communities.

References

Caputo, Darryl F. *Open Space Pays: The Socioenvironomics of Open Space Preservation,* New Jersey Conservation Foundation.

Fox, Thomas, *Urban Open Space: An Investment That Pays,* Neighborhood Open Space Coalition, New York, NY, 1990.

Frank, James E., *The Costs of Alternative Development Patterns: A Review of the Literature,* the Urban Land Institute, 1989.

Freedman, Bill, *Environmental Ecology: The Ecological Effects of Pollution, Disturbance, and Other Stresses,* Academic Press, San Diego, CA, 1995.

Hart, Maureen,

Interagency Ecosystem Management Task Force, *The Ecosystem Approach: Healthy Ecosystems and Sustainable Economies,* Report of The Interagency Ecosystem Management Task Force, Volume 1, June 1995.

Wilson, E.O., *Biodiversity,* National Academy Press, Washington, DC, 1994.

4

Land Use Management and Soil Conservation

The loss and degradation of land resources need to be seen in the context of policy, socio-economic conditions and the environment. The impact on agriculture and food production, as well as on the ecological and protective functions of natural and managed ecosystems is, however, universally recognised. Recently, the UN Millennium Declaration, the UN Millennium Development Goals and the World Summit for Sustainable Development (WSSD) Plan of Implementation recognised the maintained integrity and restoration of land resources as a critical factor in achieving economic and ecological sustainability. To meet these challenges, new and innovative approaches are required.

MDGs and Sustainable Land Use Management and Soil Conservation

In September 2000, at the United Nations Millennium Summit in New York, world leaders agreed to a set of time-bound and measurable goals and targets for combating poverty, hunger, disease, illiteracy, environmental degradation and discrimination against women. Placed at the heart of the global agenda, they are now called the Millennium Development Goals (MDGs). The Summit's Millennium Declaration also outlined a wide range of commitments to human rights, good governance and democracy. Out of the eight goals four are relevant in the context of sustainable land use management and soil conservation.

1. *Eradicate extreme poverty and hunger*
    - Target for 2015: halve the proportion of people living on less than a dollar a day and those who suffer from hunger.1.2 billion people still live on less than $1 a day. But 43 countries, with more than 60 per cent of the world's people, have already met or are on track to meet the goal of cutting hunger in half by 2015.
2. *Promote gender equality and empower women*
    - Targets for 2005 and 2015: eliminate gender disparities in primary and secondary education preferably by 2005 and at all levels by 2015.

    Two-thirds of the world's illiterates are women, and 80 per cent of its refugees are women and children. Since the 1997 Microcredit Summit, progress has been made in reaching and empowering poor women, nearly 19 million in 2000 alone.
3. *Ensure environmental sustainability*
    - Integrate the principles of sustainable development into country policies and programmes and reverse the loss of environmental resources.
    - By 2015, reduce by half the proportion of people without access to safe drinking water.
    - By 2020, achieve significant improvement in the lives of at least 100 million slum dwellers.
    - More than one billion people still lack access to safe drinking water; however, during the 1990s, nearly one billion people gained access to safe water and as many to sanitation.
    - Develop a global partnership for development
    - Develop further an open trading and financial system that includes a commitment to good governance, development and poverty reduction – nationally and internationally.
    - Address the least developed countries' special needs, and the special needs of landlocked and small island developing states.
    - Deal comprehensively with developing countries' debt problems.
    - Develop decent and productive work for youth.

- In cooperation with pharmaceutical companies, provide access to affordable essential drugs in developing countries.
- In cooperation with the private sector, make available the benefits of new technologies – especially information and communications technologies.

Too many developing countries are spending more on debt service than on social services. New aid commitments made in the first half of 2002 alone, though, will reach an additional $12 billion per year by 2006.

UNEP's LAND-RELATED ACTIVITIES

From the outset in 1972, UNEP's land-related activities have focused on medium- to long-term solutions for desertification UNEP contributed significantly to the implementation of the UN Plan of Action to Combat Desertification (UNPACD), which subsequently led to the ratification process of the United Nations Convention to Combat Desertification (UNCCD). Since then UNEP's role has gradually changed from global coordination of UNPACD to supporting the implementation of the UNCCD.

In the 1980s, UNEP developed the World Soils Policy which recognised the fact that soil is a finite resource, and that continuously increasing demands are being placed on this resource to feed, clothe, house and provide energy for a growing world population and to provide ecological balance. Governments agreed in the World Soils Policy that the use of soils should be based on the sound principles of resource management in order to enhance soil productivity, to prevent soil erosion and degradation, and to reduce the loss of good farmland to non-farm purposes.

In cooperation with the UN Food and Agriculture Organization (FAO), UNEP contributed to the World Soil Charter and assisted developing countries to formulate their national soil policies. At the international level, those instruments have contributed to raising the profile of soil conservation as a major international environmental issue.

UNEP, in cooperation with international partners, carried out global and regional assessments in the early 1990s in order to gain fast and reliable data on the global status of human-induced soil degradation. These assessments still serve as the main reference on the global extent of land degradation.

World Soil Charter

At its 12th meeting in April 1980, UNEP's Governing Council requested an active collaboration with FAO, UNESCO and other relevant international organizations to ensure preparation and adoption of a soils policy which would bring the issue of soil and land degradation to the level of international environmental policy action, acknowledging soil as a non-renewable resource.

A group of high-level experts met in Rome in 1980 to develop principles and a plan of action for soil policy, which were incorporated in the World Soil Charter.

The 21st Session of the FAO Conference, in November 1981, adopted the World Soil Charter. The World Soil Charter established a set of principles for the optimum use of the world's land resources, for the improvement of their productivity, and for their conservation for future generations.

The Charter calls for a commitment on the part of governments, international organizations and land users in general to manage the land for long-term advantage rather then for short-term expediency. Special attention is called to the need for land-use policies, which create the incentives for people to participate in soil conservation work taking into account both the technical and socio-economic elements of effective land use. The Charter provides 13 principles and guidelines for action by governments and international organisations.

Global Environmental Outlook

The UNEP Global Environmental Outlook (GEO) report is a response to the Agenda 21 request for comprehensive environmental reporting on the global state of the environment. Its third version, GEO-3, which was published in 2002, places major emphasis on providing an integrated assessment of environmental trends over the 30 years since the 1972 Stockholm conference. GEO-3's analysis of environmental trends takes into consideration the widest possible range of social, economic, political and cultural drivers and root causes.

State of the Environment 1972-2002: Land and Soil Degradation

For the last 30 years increasing pressure on land has been exerted through the growing need for higher agricultural yields to feed the increasing world population. Compared to the 1970s, 2,220 million more people need to be fed today.

The main land and soil related features in Africa and in Latin America and the Caribbean include increasing loss of agricultural area through severe land degradation as well as inappropriate and inequitable land tenure systems. In Africa particular problems include a decline in soil fertility, soil contamination, unsustainable land management and conservation, gender imbalances in land tenure, and conversion of natural habitat to agricultural or urban uses. Land degradation and desertification continue to be the most significant environmental issue in West Asia, Asia and the Pacific. For Asia and the Pacific other critical land issues include land use change and soil contamination. In Europe and North America, key issues associated with degradation of land resources include urban sprawl, soil contamination and erosion. The Polar regions face serious land degradation in the form of soil erosion through resource exploitation and infrastructure measures.

GEO Outlook 2002-32: Four Scenarios

GEO-3 also attempts a look into the next 30 years by applying a forward-looking and integrated analysis, which is based on four basic scenarios:

- *Markets First:* A market-oriented approach adopting the values and expectations prevailing in today's industrialized countries.
- *Policy First*: Government initiatives like policy measures, regulatory frameworks and planning processes prevail.
- *Security First*: A world of striking disparities where inequality and conflict prevail.
- *Sustainability First*: A visionary state of affairs with new, more equitable values and institutions, which supports sustainability.

With respect to land degradation the different scenarios vary in their impact. In Africa, the risk of increased land degradation and desertification is omnipresent. In the Policy First and Sustainability First scenarios, easier access to support services will help farmers to manage land resources better and policies based on integrated land management become commonplace in the region. At the other end of the spectrum, in a Security First scenario the high concentration of people elsewhere is expected to contribute to severe land degradation and soil erosion, while reasonable conditions are maintained in the protected areas serving the land-owning elite. Similar problems arise in the Markets First scenario as more, better quality agricultural land will be taken over for commodity and cash crop production.

Land and forest degradation as well as forest fragmentation are expected to remain among the most relevant environmental issues in Latin America and the Caribbean in all scenarios. Significant loss of forest area may occur in a Markets First scenario. In a Security First world, the control over forest resources by transnational companies that create cartels in association with the national groups in power, might promote the growth of some forest areas, but this is not thought to stop net deforestation. More effective management can ameliorate some of these problems under a Policy First scenario. Unsound deforestation stops almost completely in a world of Sustainability First.

KEY ISSUES

Land and Soil Degradation

The UN Millennium Declaration further states that "nearly 2 billion hectares of land, an area about the combined size of Canada and the United States, is affected by human-induced degradation of soils, putting the livelihoods of nearly one billion people at risk. [...] Each year an additional 20 million hectares of agricultural land either becomes too degraded for crop production, or becomes lost to urban sprawl"

Poverty

Land degradation is both a cause and an effect of poverty. The spiral of impoverishment and environmental decline is driven by external factors, namely commercialisation, civil strife, displacement and natural hazards, and internal factors, such as population growth, governance and existing poverty People in marginal lands are especially vulnerable. General perceptions on the impact of environmental degradation on poor people conclude that: a) poor people are more vulnerable to loss of biological resources, b) extreme environmental stress can force the poorest to migrate, c) inequality reinforces environmental pressure, and d) global and national policies can create or reinforce a vicious cycle of poverty /environmental degradation

Land Tenure and Public Participation

Inadequate care of land resources, and related poverty, is often directly linked to issues of land tenure and in turn impacts on the use or over-use of these resources. While secure land ownership and rights per se do not always provide a guarantee for sustainable land use management they are, however, a necessary prerequisite. Unsustainable practices of land resource management are directly related to the level of awareness and, consequently, to the level of public participation in decision making on environmental issues. Awareness raising, education and training provide an important but often missing link in the mitigation and control of land degradation.

Environmental Impact of Agriculture

The main pressure on land resources is to increase food production for a growing global population' / despite unprecedented growth in agricultural productivity over the past three decades On a global average, current food production should satisfy the world's population as many regions have a surplus of agricultural produce through intensified industrial production. However, strong regional differences occur, with some regions being net importers of agricultural goods, and requiring further extension of arable land The intensification of agricultural production lessens the need for reclamation of natural areas, but often requires an increased use of herbicides, pesticides, and fertilisers resulting in a decline of environmental quality and biodiversity The impact of land use and soil management techniques on soil biodiversity and its functioning is of increasing importance. Pesticides may cause health hazards if applied and disposed of inappropriately Unregulated fertiliser input, often subsidised, causes water pollution, biodiversity shifts and health threats. The potential impact of the increasing use of genetically modified plants on biosafety is, as yet, largely unknown. The extension of farming into barely productive, marginal lands through, for example the conversion of forests, wetlands, and mountain slopes causes rapid fertility decline featuring erosion and nutrient loss. It affects environmental services such as water supply and micro climate, and often creates related hazards such as land slides or desertification.

Water and Land Use Management

Agriculture is responsible for about 70% of all freshwater withdrawals, of which 70% is wasted. Inefficient irrigation schemes lead to unsustainable water logging and irreparable salinisation and alkalisation of soils, especially in the case of groundwater withdrawal 15 Estimates of up to 80% of global marine and freshwater pollution is derived from land-based activities such as contamination from municipal and industrial waste and agricultural fertilisers and pesticides. Water erosion and sedimentation further aggravate the situation, as does the loss of productive wetlands and floodplains and the anticipated global changes in climatic and hydrological patterns

Environmental Emergencies and Land Use

Unsustainable land use practices, especially on marginally productive lands, or the increasing use of unsuitable and unsafe lands such as steep slopes and river banks, are a major factor in the increasing frequency and severity of certain natural disaster types. These can have severe impacts on people and ecosystems. People depending on marginal lands are more vulnerable to the effects of disasters such as hurricanes, floods, droughts, earthquakes, and land slides.

Urbanization

Urbanization, including infrastructure development, causes an increasing loss of limited natural resources and habitats, soil sealing, slope instability, erosion and river siltation, diffuse and local soil and groundwater contamination through industrial waste, chemicals and air pollutants. Urban agriculture, though an essential aspect of urban life in many developing country economies, providing food and income, causes additional air, water and soil pollution from improper use of fertilisers and pesticides The urban poor, who cannot afford expensive remediation or a move to cleaner areas, suffer the most from loss of resources and health threats.

Global Climate Change

Natural systems can be especially vulnerable to climate change and some of these systems may undergo significant and irreversible damage. Also, many human systems are sensitive to climate change and some are vulnerable Impacts will depend on the adaptive capacity of natural systems and the resource availability of societies. For non-irrigated agriculture in drylands, yield declines of as much as 30% are expected during this century. In turn, land use change has direct and significant implications on the global carbon cycle Land use change, mainly deforestation, accounts for approximately 33% of all global anthropogenic carbon emissions over the past 150 years As land degradation almost always implies a loss of carbon, much attention is given to land use management options that restore organic matter and soil fertility through carbon sequestration.

Trade and Environmental Externalities

Trade liberalisation and trade related policies often bring about a tendency to concentrate on the increase of economic returns and may consequently lead to overexploitation of water and nutrient resources and the dismissal of sustainable practices such as fallow periods and crop rotation. However, trade liberalisation and trade related policies may also imply positive effects on the environment by introducing environmentally friendly technologies or policies Generally, the degradation of land resources involves direct and indirect costs. Environmental externalities and long-term implications of land degradation are often more severe than direct costs of forgone income Inefficient governance and capacity deficiencies in many developing countries are equally important factors contributing to increasing land degradation.

Challenges

The prevention and mitigation of land degradation through the promotion of sustainable land management is a global challenge. To address the identified key issues in land use management and soil conservation within a development oriented approach poses challenges to all stakeholders and requires integrative solutions across the policy, socio-economic, and environment sectors. The relevant framework for tackling these challenges is set inter alia in Agenda 21, the UN Millennium

Declaration and the WSSD Plan of Implementation

Devastation - the removal of vegetation in Western Australia' s farming areas has allowed groundwater levels to rise. In many places soil salination has resulted, and productive land is being lost to salt at a rate of 18,000 hectares a year.

UNEP in particular, is challenged to address the environmental dimensions of land use management and soil conservation as they relate to the overall objectives of sustainable development and poverty reduction. UNEP is further challenged to support governments and civil society in achieving environmentally sustainable land use. Consequently, UNEP is to develop and apply environmentally focused and development orientated policy guidance in close cooperation with governments, civil society and fellow UN and international organisations. UNEP must also work with all of these entities to ensure that laws at the international, regional and national level that govern land use management and soil conservation are fully implemented.

UNEP's Goals to Sustainable Development

UNEP's ultimate mandate is to contribute to sustainable development and poverty reduction by focusing on specific environmental dimensions. Based on the identified key issues and challenges as well as on UNEP's expertise and its renewed mandate as in the Nairobi Declaration /, UNEP's primary goals with regard to land use management and soil conservation are identified as follows:

a. An ecosystem approach for land use management and soil conservation applied and interlinkages and synergies within and across relevant sectors developed;

b. A global land cover monitoring process and assessment of the state of land resources in partnership with other UN organisations and partners developed and implemented;

c. Environment focused and development orientated policies on sustainable land use management and soil conservation developed and implemented. To be achieved through capacity building, information management and public participation, response to environmental emergencies, development of legal instruments, regional co-operation and the development, implementation and execution of GEF projects;

d. Cooperation with scientific centres of excellence extended in order to strengthen science-policy interaction and knowledge systems through partnerships with governments and civil society; and

e. Further support to the implementation of the UN Convention to Combat Desertification and specific support to Africa in regard to land degradation through the NEPAD Environment Initiative.

The identified goals are in line with the action areas in the WSSD-Workinggroup on Water, Environment, Health, Agriculture, and Biodiversity (WEHAB) Framework of Action on Agriculture, namely to a) increase agricultural productivity and sustain the natural resource base contributing to efforts to eradicate poverty and ensure environmental sustainability, b) encourage knowledge generation, c) establish innovative public-private partnerships to stimulate joint implementation of sustainable agriculture and natural resource conservation, and d) develop enabling policies, associated institutional reforms and regulatory frameworks. Equally, the identified goals are in line with the WEHAB Frameworks of Action on Water and Biodiversity.

WEHAB Initiative

The WEHAB initiative was launched by UN Secretary-General Kofi Annan as a contribution to the World Summit on Sustainable Development (WSSD). WEHAB seeks to provide focus and impetus to action in the five key thematic areas of Water, Energy, Health, Agriculture and Biodiversity. These key areas are integral to a coherent implementation of sustainable development and are among the issues reflected in the Summit's Plan of Implementation.

WEHAB's Framework for Action on Agriculture focuses a new spotlight on the role agriculture, agro-biodiversity, and integrated natural resource management can play in national, regional and global development sustainability. At the same time it is acknowledged that for fostering environmental sustainability, agriculture's large and growing ecological footprints need to be reduced.

Ecosystem Approach

An ecosystem is a dynamic complex of plant, animal and micro-organism communities and their non-living environment interacting as a functional unit. An ecosystem can refer to any functioning unit at any scale: it could be a grain of soil, a forest, a biome or the entire biosphere.

The Ecosystem Approach relates to relevant levels of biological organization, which encompass the essential structure, processes, functions and interactions among organisms and their environments. It recognizes that humans, with their cultural diversity, are an integral component of many ecosystems.

The ecosystem approach requires holistic adaptive management to deal with the complex and dynamic nature of ecosystems and the complete knowledge or understanding of their functioning. It can integrate other management and conservation approaches and other methodologies to deal with complex situations. There is no single way to implement the ecosystem approach, as it depends on local, provincial, national, regional or global conditions.

Strategies and Intended Action

Ecosystem Approach for Land Use Management and Soil Conservation

The ecosystem approach is focused on the integrated management of land, water and living resources and promotes conservation and sustainable use of resources in an equitable way It is tailored to reflect the various aspects of land use management and soil conservation in a functional, cross-sectoral and integrative manner underlining the ecological and socio-economic functioning of land resources Through the ecosystem approach a direct link is made between environmental land and soil issues and sustainable development and poverty reduction. As the ecosystem approach has been developed from an environmental perspective it is important to identify the degree of conformity with other, more 'productivity' orientated concepts

Applying the ecosystem approach to environmental land and soil issues requires establishing stronger links to other relevant UNEP focal areas as defined in Governing Council decisions, and strengthening land and soil components in context:

Environmental Land and Soil Issues as Relevant to UNEP's Portfolio

Many focal areas within UNEP's portfolio relate directly or indirectly to land degradation, land use management and soil conservation. Direct links exist to Governing Council decisions on forest-related issues /, on chemicals /, on water /, on climate change /, and on trade Land and soil issues are also reflected in UNEP's activities with regard to support to Africa and others /, environmental emergency prevention /, International Environmental Governance (IEG) /, and the work of the Global Environment Facility (GEF)

The majority of multilateral environmental agreements (MEAs) relate either directly or indirectly to land and soil issues Agreements of direct relevance include inter alia the United Nations Convention to Combat Desertification (UNCCD), the African Convention on the Conservation of Nature and Natural Resources, the ASEAN Agreement on the Conservation of Nature and Natural Resources, and the Alpine Convention Soil Protection Protocol. Both direct and indirect links to land and soil issues are found within the biodiversity related MEAs, including inter alia the Convention on Biological Diversity (CBD) and the Ramsar Convention on Wetlands /, the chemistry related MEAs, including inter alia the Stockholm Convention on Persistent Organic Pollutants, and the atmosphere related MEAs, including inter alia the UN Framework Convention on Climate Change (UNFCCC), the Kyoto Protocol and the Convention on Long-Range Transboundary Air Pollution.

Identifying such synergies at all levels of UNEP's core areas including environmental assessment, policy development and implementation contributes to the efficient and coherent implementation of multilateral environmental agreements.

Intended action:

i) To strengthen the coherent integration of land and soil issues within and across relevant focal areas and MEAs at all levels of UNEP's core activities,

ii) To support governments and civil society in strengthening the integration of land and soil issues within and across the implementation of MEAs.

Relating land and soil issues to other environmental focal areas

A complementary UNEP land and water policy

Integrated land and water management is a key principle of successful water management. This is due to the many interactions between land use and water management. Environmentally sustainable land use management and soil conservation is essential for achieving environmentally sustainable water use in terms of both quantity and quality. Effective frameworks such as Integrated Water Resource Management (IWRM) already exist and need to be built upon Activities for the implementation of UNEP's Water Policy and Strategy are highly relevant to land-related issues It is therefore necessary that complementary land and water activities be developed and implemented, particularly in relation to the Global Programme of Action for the Protection of the Marine Environment from Land-based Activities, to freshwater management and to relevant regional processes

Intended action:

iii) To develop practical policy and management guidance on integrated land and water management following existing approaches such as IWRM,

iv) To develop joint land-water initiatives for integrated and coherent national and transboundary assessment, policy development and implementation on land use and water management.

Land use Management, Biodiversity and Forests

The integration of the sustainable use and conservation of biodiversity is among the most prominent challenges for biodiversity policies and the land use sector This is particularly true for drylands, which are well known for their genetic diversity within species, rather than for species variation or "species richness".

The various functions of agriculture including income generation, food and fibre production, and environmental benefits such as provision of biodiversity, water, landscaping, recreation and so forth, require the further identification and development of policies and instruments that support an integrated approach across such functions The links between biodiversity and alternative management options such as organic agricultureare of interest in this context.

Land use includes forests. UNEP' s policy action on forest-related issues includes inter alia the relationship between forest management, deforestation and biodiversity /, the promotion of protected forest areas, the promotion of common issues and needs of low forest cover countries, trade and environment in relation to forest products and services, and tradeoffs between forest production and environmental services.

Intended action:

v) To promote integrated management of and equitable access to biodiversity, and to sustain biological diversity in agriculture and drylands by developing policy guidance and supporting implementation as based on relevant decisions by CBD, UNCCD and other multilateral environmental agreements,

vi) To support the implementation of the Intergovernmental Panel on Forests (IPF) and the Intergovernmental Forum on Forests (IFF) proposals for action on sustainable forest management and to strengthen UNEP's role in the implementation of the UN Forum on Forests (UNFF) and the Collaborative Partnership of Forests (CPF) in the context of the WSSD Plan of Implementation.

Land Use Management and Climate Change

The implementation of the UNFCCC and the climate change related sections in the WEHAB Framework of Action / require the development of concepts which identify, evaluate and interrelate vulnerability, adaptation and mitigation measures in natural resource management. Vulnerability assessments and related capacity building, especially in regions and sectors most likely to be affected, are prerequisites for effective adaptation measures. The efficient mainstreaming of adaptation concepts into sectoral planning processes requires a focus on the identification of no-regret options and synergies with objectives of sustainable development and poverty reduction

Terrestrial carbon sequestration is one option for the mitigation of greenhouse gas emissions The implementation of land use, land use change and forestry related (LULUCF) activities in the context of the Kyoto Protocol provides both challenges and opportunities. A strong enabling context at the national and international level will be required to implement environmentally sound and socially equitable climate change mitigation projects in the land use and forestry sectors Also, synergies between mitigation and adaptation measures in the land use and forestry sectors are likely to provide opportunities for improved concepts in addressing the anticipated impacts of global climate change.

Intended action:

vii) To assess the vulnerability of land resources in relation to climate change and climate variability and to develop and support the implementation of effective adaptation measures in the context of sectoral planning processes

viii) To develop and support the implementation of frameworks for environmentally integrated, economically viable and socially equitable carbon sequestration concepts by strengthening links to objectives in, for example, dryland management, biodiversity conservation, forest landscape restoration, water use management and rural livelihood development.

Land and Soils as Impacted by Chemicals, Industrial Waste and Urbanization

Increasingly, a significant proportion of the decline in land resource quality is caused by industrial waste, mining, diffuse and spot contamination involving heavy metals, acids, pesticides, herbicides et al. as well as soil sealing through urban sprawl Additionally, the environmental impacts from the agri-food production and consumption cycle require international and national assessment and response. UNEP is well positioned to encourage technologies, practices and behaviour that are less polluting and make more efficient use of natural resources.

Intended action:

ix) To assess the environmental impacts throughout the agri-food chain including the trade of agricultural products and to develop responses to key issues through MEAs, information exchange, policy guidance on and implementation of remedy options.

Land Use Policy, Trade and Poverty Reduction

Assessments of the static and dynamic interlinkage between different economic sectors, different economic agents and the environment are important in order to identify the impact of trade liberalisation and trade-related policies on the environment. The integrated treatment of policies underpinning economic development may do much to enhance the viability of policies directed towards the environmental and social aspects of sustainable development. Measures identified to alleviate environmental, economic and social problems include a mix of sector-specific policies, broader macro-economic policies and environmental policies.

The global carbon cycle, showing the carbon stocks in reservoirs (in Gt C=10) and carbon flows (in Gt C yr) relevant to the anthropogenic perturbation as annual averages, 1989 – 1998 (Schimel et al.1996; Watson et al. 2000). Terrestrial ecosystems play an important role in the global carbon cycle. Around one-third of global anthropogenic carbon emissions in the past 150 years resulted from land-use change, namely forest clearing in the tropics and elsewhere. On an annual global basis, land-use change results in emissions of 1.6 ± 0.8 gigatonnes of carbon (Gt C). This accounts for around 25 percent of emissions from fuel combustion and cement production. These and other findings were assessed by the UNE/World Meteorological Organisation (WMO) Intergovernmental Panel on Climate Change (IPCC).

Land resources contribute to human wellbeing in many ways and must not be seen as a mere commodity. The ecosystem services that are provided by land resources include a) provision of the production base for food, fibre etc, b) regulation of the fluxes of water, nutrients and other substances and c) enrichment by providing cultural and religious services While all people depend on these services, poor people are more heavily dependent on them. The environmental integrity of land use management and soil conservation can only be ensured if these services of land resources are balanced. However, as these functions often represent the non-market sector, they are as yet underestimated and undervalued in their importance for poverty reduction.

Intended action:

x) To explore positive and pro-active commitments between trade and sustainable land resource management,

xi) To identify, evaluate and integrate the ecosystem services of land resources in the UNEP poverty-environment nexus.

Integrated Water Resource Management

One of the challenges of water management is to accelerate the shift from mono-dimensional water managing water approaches – such as water for cities or water for agriculture – to integrated water resource management in order to meet a variety of needs.

Increased water and chemical use in irrigated agriculture has benefited farmers and the poor. But increased water and chemical use that fueled the Green Revolution has also contributed to environmental degradation, and threatened the resource base upon which we depend for food and livelihoods.

Rapid deterioration of water quality in many areas poses a serious threat to the sustainability and the safety of food production systems. Pollutant loads have increased enormously due to increased industrial, urban and agricultural uses. Degradation of dry lands is an urgent global problem placing people mainly in developing countries at risk. In highly industrialized regions ameliorating soil contamination and combating acidification are priorities. Millions die every year as a result of diseases caused by contaminated water.

Integrated management of water for hydropower, domestic use, agriculture and for nature demands the identification of environmental requirements in a first step. The second step is to match the temporal quality, quantity and demand patterns of various sectors. It is insufficient and unwise to allocate an annual quantity of water to the environment. Natural systems have demands for water that vary enormously in time and space. In fact, floods and droughts may be more beneficial than average flows, which are highly prized in agriculture. Recognizing these temporal requirements, it should be possible to allocate water to nature, and distribute it in a manner when its value for nature is highest, while at the same time meeting the demands of agriculture.

A classical IWRM framework contains a national water policy, strategies and legislation; an information system; allocation scenarios and action plans, either at the national or basin level; co-ordination, financing and monitoring mechanisms to implement the plans; governance mechanisms to ensure transparency and accountability; and a unique organization responsible for the whole thing. Without such a framework, pollution and disputes over limited and vulnerable water resources will continue to develop between rural, industrial and urban users, with the aguatic environment and human development continuing to suffer as a result.

Conserving Below-ground biodiversity

Under the co-ordination of the Tropical Soil Biology and Fertility Institute of CIAT a consortium of research institutes from seven countries (Brasil, Mexico, Cote d'Ivoire, Uganda, Kenya, India and Indonesia) are working together with the wide spectrum of stakeholders to develop methods for valuing below-ground biodiversity and its services and identify the necessary conservation actions.

In order to allow these countries to address this challenge the Global Environment Facilities through the United Nations Environment Program is supporting a five-year project. The objective of this project is to enhance awareness, knowledge and understanding of below-ground biological diversity (BGBD) important to sustainable agricultural production in tropical landscapes by the demonstration of methods for conservation and sustainable management. The project will explore the hypothesis that, by appropriate management of above- and below-ground biota, optimal conservation of biodiversity for national and global benefits can be achieved in mosaics of land-uses at differing intensities of management and furthermore result in simultaneous gains in sustainable agricultural production.

Expected Outputs

1. Internationally accepted standard methods for characterization and evaluation of BGBD, including a set of indicators for BGBD loss.
2a. Inventory and evaluation of BGBD in benchmark sites representing a range of globally significant ecosystems and land uses.
2b. A global information exchange network for BGBD.

3. Sustainable and replicable management practices for BGBD conservation identified and implemented in pilot demonstration sites in representative tropical forest landscapes in seven countries.
4. Recommendations of alternative land use practices and an advisory support system for policies that will enhance the conservation of BGBD.
5. Improved capacity of all relevant institutions and stakeholders to implement conservation management of BGBD in a sustainable and efficient manner.

Agri-food Production and Consumption

The whole agri-food chain, which involves the production of agricultural products, of food transformation, food distribution by the retail chain, and consumption have major environmental impacts. UNEP is addressing these issues through its main work programmes e.g. DTIE, in co-operation with other relevant UN Agencies.

Depletion and contamination of natural resource occurs throughout the agri-food chain. Pollution and food contamination related to the use of production technologies and processes, as well as from the use of the products aimed at increasing agricultural yields and facilitating food conservation, also have important environmental consequences. Bio-safety and food safety are as well important issues.

Key environmental impacts in agri-food production include depletion of natural resources like water, fisheries, forests and the loss of bio-diversity, land degradation resulting from unsustainable agricultural practices and from deforestation, land contamination and possible crop contamination resulting from uncontrolled use of fertilizers and pesticides contaminated with heavy metals, ground water contamination by nitrates and pesticides resulting from uncontrolled use of fertilizers and pesticides and intensive life-stock production, CO2 and air emissions stemming from energy use throughout the agri-food chain, hazardous and urban waste generated from intermediary products e.g. containers for holding pesticides and food packaging, and food and bio-safety issues related to the increased use of land for agriculture and the use of genetically modified organisms.

UNEP responds to the impacts through better assessing and understanding of environmental impacts throughout the agri-food chain including the trade of agricultural products, through developing responses to key issues through international environmental agreements, voluntary initiatives, policy guidance, and information exchange, through assisting in the implementation of those responses, and through monitoring and evaluating of trends.

Global Land Cover Monitoring and Assessment

A reliable assessment of the status and trends of global land cover is a prerequisite for adequate environmental policy development and implementation. It requires a scientifically qualified global land cover monitoring process in cooperation with competent partners including FAO and others. Key issues are to be integrated into environment assessment and early warning processes, in particular UNEP's Global Environment Outlook (GEO).

UNEP's assessment strategy supports the development and strengthening of regional and national capacities for collection, harmonisation, analysis and reporting of land and soil data as a base for a coherent global assessment system. The strategy further includes development of improved access to land assessment products and information.

Integrated land and soil assessments require further development of cross-sectoral, science-based indicators /, especially as growing evidence shows that current concepts are partly misleading, resulting in ineffective remedy policy concepts

Intended action

— To develop and implement a global decadal land cover monitoring process focusing on environmentally sensitive areas and the strengthening of national and regional capacities for environmental assessment data and information management.

Policy Development and Implementation

Policy development and guidance to prevent and mitigate the environmental and social impacts of land degradation requires: a) the identification of constraints and barriers in policy, administration and culture, b) creation of an enabling environment, including capacity building and institutional arrangements for participatory public-private partnerships, c) creation and access to public information systems and d) the provision of technical support to governments and civil society for decision-making and e) to mainstream land and soil related issues into development policies.

Supporting national and regional legal processes and structures for the integration of the environmental dimension of land use management and soil conservation is a key component of policy development. UNEP's third Montevideo Programme has as its objective support to governments in improving the conservation, rehabilitation and sustainable use of soils by promoting the development and implementation of laws and policies for enhancing the conservation, sustainable use and, where appropriate, the rehabilitation of soils.

Policy implementation focuses on several issues including: a) pilot project development, b) capacity building in cooperation with governments and civil society, c) response to environmental emergencies through preparedness, prevention, mitigation and response, and d) cause analysis and identification of possible policy implications at national and global level, e) development of tools and guidelines, and f) awareness raising, education and training through, for example, UNEP's Best Practices and Success Stories Global Network (BSGN).

Regional cooperation is essential for relating policy development and implementation into governmental and intergovernmental dialogue. Additionally, regional cooperation links priorities with respect to land and soil issues within regional ministerial fora, existing regional networks and centres of excellence. Regional and sub-regional cooperation will further enhance the need for trans-boundary land resource management and emergency strategies. It will catalyse support for the implementation of multilateral environmental agreements (MEAs), namely the UNCCD.

The GEF operational programme on sustainable land management aims to mitigate the root causes and negative impacts of land degradation on terrestrial and aquatic ecological systems through sustainable land management Key elements of UNEP's GEF strategy on land degradation include: a) assessments of land degradation to better evaluate the interlinkages between land degradation and other GEF focal areas, b) development of tools and methodologies for sustainable land management /, c) targeted research focusing on developing models for the sustainable use of ecosystems within the managed landscape, d) management of transboundary land and water resources including the implementation of UNCCD Regional and Subregional Action Programmes and other regional frameworks, and e) development of capacity for natural disaster preparedness focusing on mitigation of land degradation caused by drought and flooding.

Intended action

— To prevent and mitigate the environmental and social impacts of land degradation through policy guidance, capacity building, response to environmental emergencies and regional cooperation,

— To support the development and implementation of legal instruments for national and multilateral integration of environmental aspects of land use management and soil protection,

— To support governments and partners in the development, implementation and execution of GEF projects, in particular with reference to the GEF operational programme on sustainable land management and other related GEF operational programmes and in reflection of internal expertise.

Land Degradation Control Initiative by UNEP

For more than 20 years, UNEP has been actively involved in worldwide efforts to combat dryland degradation and until the adoption and entry into force of the UNCCD in 1998, had a special programme, the Desertification Control Programme Activity Centre (DCPAC), promoting and supporting desertification control initiatives globally, regionally and even at national levels through assessment, awareness rising and institutional capacity-building. It was under DCPAC that the Success Stories in Desertification / Land Degradation Control Initiative was implemented in 1994.

The initiative aims at publicising successful projects and community-based initiatives in desertification control to raise global awareness that land degradation in the drylands can be both prevented and corrected, as well as to build confidence in community responsibility for the local environment and in local abilities to solve land management problems.

The UNEP success stories initiative is helping to develop capacity through the replication of best practices. It is a global programme coordinated from UNEP headquarters, but implemented in close collaboration with UNEP Regional Offices, NGOs and civil society in the regions.

Between 1995 and 1999, 25 case studies evaluated around the World won the UNEP "Saving the Drylands Award". These success stories from Africa, Asia, and Latin America and the Caribbean address not only the biophysical but also the socio-cultural-economic issues in all its development stages, thus ensuring long-term sustainability.

GEF Operational Programme #15

The main objective of OP #15 is to mitigate the causes and negative impacts of land degradation on ecosystem stability, functions and services through sustainable land management practices as a contribution to improve people's livelihoods and economic wellbeing. Within OP#15 GEF support to sustainable land management activities can be provided with regard to:

Capacity building

— Mainstreaming of sustainable land management into national development priorities
— Integration of land use planning systems
— Agreements and mechanisms for management of transboundary resources

On-the-ground investments

— Sustainable agriculture
— Sustainable rangeland/pasture management
— Forest and woodland management

Targeted research

— Better understand the policy and institutional failures that drive land degradation.
— Facilitate the refinement and adoption of innovative Sustainable Land Management practices and technologies, including early warning and monitoring systems.
— Some key GEF projects in the area of land use management and soil conservation as implemented by UNEP.

Desert Margins Programme (DMP)

The overall objective of the DMP is to arrest land degradation in Africa's desert margins through demonstration and capacity building activities. The

GEF increments to this project will enable the programme to address issues of global environmental importance, in addition to the issues of national economic and environmental importance, and in particular the loss of biological diversity, reduced sequestration of carbon, and increased soil erosion and sedimentation. The project will make a significant contribution to reducing land degradation in the marginal areas and help conserve biodiversity. Guidelines and recommendations domains and supportive national policies that address biodiversity concerns will be set in place in implementing countries.

Land Use Change Analysis

The project goal is to contribute to the conservation of biodiversity and prevention of land degradation by providing useful instruments to identify and monitor changes in the landscape associated with biodiversity loss and land degradation, and identify the root causes of those changes. These tools will assist GEF in the design of multi-focal area projects. With information obtained from such tools, stakeholders and decisionmakers will be better able to implement effective remedial and preventive policy.

Land Degradation Assessment in Drylands (LADA) Project

The principal objectives of the LADA project are two-fold. First, to develop and implement strategies, tools and methods to assess and quantify the nature, extent, severity and impacts of land degradation on ecosystems, watersheds and river basins, and carbon storage in drylands at a range of spatial and temporal scales.

The assessment will integrate biophysical factors and socio-economic driving forces. Second, the project will build national, regional and global assessment capacities to enable the design and planning of interventions to mitigate land degradation and establish sustainable land use and management practices. These objectives are expected to overcome current policy and institutional barriers to sustainable land use in dryland zones and establish incentives to promote the accrual of global biodiversity benefits at national and local levels.

Improved Science-policy Interaction and Knowledge Systems

Given its specific mandate and expertise, UNEP, as the global environmental body of the UN, plays a complementary role within the UN system in addressing the environmental aspects of land and soil policies in relation to sustainable development Existing processes including the Environment Management Group (EMG) and UNEP's IEG process provide appropriate channels to ensure efficient and coherent strategies in cooperation with governments and fellow UN organisations.

Improved science-policy interaction is required in order to strengthen and extend knowledge systems as outlined in the WSSD Plan of Implementation and WEHAB framework for action UNEP can achieve this by strengthening and extending its partnerships with scientific centres of excellence in the area of land use management and soil conservation. Continued compilation and dissemination of information on best practices in land use management, including the development of databases, is another important component in supporting policy implementation.

Strengthening cooperation with civil society and enhancing public-private partnership is at the core of UNEP's mandate. UNEP's strategy on enhancing civil society engagement in its work stresses the importance of close cooperation with civil society for substantive input into, and ownership and implementation of, environmental policy Modalities for civil society's input entail capacity building and a continuous dialogue on key issues including land use management, environmental services and poverty reduction.

Intended action

— To strengthen the cooperation with scientific centres of excellence and governments on the environmental aspects of land use management and soil conservation,

— To develop partnerships with civil society in order to enable dialogue and capacity building in relation to environmental aspects of land and soil policy development and implementation.

Improved science-policy interaction is required in order to strengthen and extend knowledge systems as outlined in the WSSD Plan of Implementation and WEHAB framework for action.

UNEP can achieve this by strengthening and extending its partnerships with scientific centres of excellence in the area of land use management and soil conservation.

Implementation of the UN Convention to Combat Desertification and Support to Africa

The UNCCD highlights the fact that "desertification is both a primary cause and a consequence in the environment-poverty nexus". The UNCCD recognises that the "harmonisation of multilateral environmental agreements and their effective inclusion into poverty reduction strategies" is required for successfully integrating policies .

In this context UNEP supports the implementation of the UNCCD on a global, regional and sub-regional level, in particular in relation to: a) major assessment processes; b) survey and evaluation of existing networks, institutions, agencies and bodies for information on and implementation of the UNCC; c) implementation of UNCCD regional and subregional action plans; Max-Planck Institute of Meteorology model ECHAM4 for 2080: Country-level climate change impacts on rain-fed cereal production potential on currently cultivated land. Source: Fischer et al. 2002. d) linkages between scientific and technical advisory processes of conventions and other MEAs for improved information management also within the UNEP-led process on IEG.

Support to Africa is one of UNEP's five areas of intervention. Therefore, UNEP strongly supports the Environment Initiative of the New Partnership for Africa's Development (NEPAD) An identified first priority area for intervention of the NEPAD Environment Initiative is to combat desertification, drought and land degradation Based on existing initiatives and existing efforts, activities required by the NEPAD Environment Initiative include a) the establishment of a regional network of centres of excellence, b) the dissemination of information on best practices, c) and strengthening the mobilisation of scientific, technical and institutional capacity building for integrated sustainable land management and d) development of regional land use guidelines and policy frameworks.

Intended action

- To continue supporting the implementation of the UNCCD, in particular assessment, national and regional action plans, information networking and best practices, and scientific and technical linkages to the CBD, UNFCCC and other MEAs,
- To assist African governments in developing and implementing stated priorities within the NEPAD Environment Initiative, with particular

reference to combating desertification and addressing land degradation issues.

UNEP and Civil Society

The 1972 Stockholm Conference on the Human Environment, which led to the founding of UNEP, was the result of the unprecedented role played by NGOs in shaping the global environmental agenda. Agenda 21 strongly advocated the need for new forms of participation in support of common efforts for sustainable development. Successive UNEP Governing Council decisions and World Summits have emphasized the need for working with the widest possible range of public organizations.

In 2000, the Malmö Ministerial Declaration reaffirmed the critically important role played by civil society in addressing environmental issues. It highlighted the need for national governments and for UNEP and international organizations to enhance the engagement of civil society organizations in their work on environmental matters.

The challenge for UNEP and its partners is to mainstream civil society participation in UNEP' s global governance and programmes of action.

For this reason, UNEP developed a "Strategy paper on enhancing civil society engagement in UNEP. This strategy has now and been finalized and was presented to the 22 Governing Council, in Nairobi, Kenya, February 2003. The strategy is based on three pillars:

1. Strengthening institutional management: to facilitate transparent and meaningful communication between civil society and UNEP, especially through internet-based technologies. This includes the development of environment-related databases, networks and environmental knowledge centres in the regions.
2. Engagement at the policy level: to involve civil society expertise and views on the development of UNEP' s work programme and discuss emerging environmental issues.
3. Engagement at the programmatic level: to cooperate with civil society on the implementation of UNEP' s work programme.

The Environment Initiative of NEPAD

The New Partnership for Africa's Development (NEPAD) adopted by the African Heads of State and Government as a framework for sustainable development in Africa calls for a coherent action plan and strategies to address the region's environmental challenges while at the same time combating poverty and promoting socio-economic development. In response, an action plan for the environment initiative of NEPAD covering the first decade of the twenty-first century has been prepared through a consultative and participatory process under the leadership of the African Ministerial Conference on Environment (AMCEN) and in close cooperation with the Secretariat of NEPAD and the African Union as well as with the support of the United Nations Environment Programme (UNEP) and the Global Environment Facility (GEF).

The action plan for the environment initiative of NEPAD focuses on various programmatic areas, such as: combating land degradation, drought and desertification; conserving Africa's wetlands; preventing, control and management of invasive alien species; conservation and sustainable use of marine, coastal and freshwater resources; combating climate change; cross-border collaboration and management of natural resources including forests, biodiversity, plant genetic resources and cross-cutting issues of health and environment, poverty and environment, assessment and early warning for natural disasters as well as biosafety issues. An integral component of the action plan is a capacity building programme for its effective implementation.

The action plan provides an appropriate framework for the establishment of a strong partnership for the protection of the environment between Africa and its partners based on the commitments contained in the United Nations Millennium Declaration.

Mechanisms for Implementation and Resources

Under its functional approach and within its institutional structure, UNEP has already embarked on enhancing its cross-divisional cooperation. Mechanisms for a cross-divisional and functional implementation of the land and soil strategy require internal identification, development and evaluation of flexible, efficient and effective structures and processes. Additionally, regular reviews on progress toward the stated goals are required as an integral part of the implementation mechanism.

The mobilisation of additional financial, institutional and human resources is crucial for the implementation of UNEP's land and soil strategy. This is particularly in view of the overall contribution to the WSSD Plan of

Implementation and the WEHAB Framework for Action on Agriculture, Water and Biodiversity. UNEP's functional approach can ensure a more cost-efficient development and implementation of policies. The integration of sponsoring governments and, increasingly, the private sector in the early stages of programme and project development through improved information exchange is crucial. The WSSD Plan of Implementation and WEHAB Framework of Action underline the importance of resource mobilisation to achieve jointly agreed goals.

Intended Action

— To strengthen cross-divisional coherence on land and soil issues for efficient implementation of the strategy within UNEP's institutional structure,

— To mobilise additional resources for the implementation of UNEP's land and soil strategy through specific partnerships with governments and the private sector.

UNEP's functional approach can ensure a more cost-efficient development and implementation of policies.

The key environmental issues as related to land use management and soil conservation are complex. They range from land use change and unsustainable management to industrial pollutants, decreasing agricultural productivity, contamination of marine ecosystems and health threats. Each of these issues has a strong policy, socio-economic and environment component.

UNEP is challenged to relate environmental aspects of land use management and soil conservation to the objectives of sustainable development, in particular poverty reduction. UNEP' s expertise in environmental assessment, policy guidance and implementation is key for an improved integration of environmental land and soil aspects across other environmental focal areas and in international, regional and national development processes, in particular the UN Millennium Development Goals and the WSSD Plan of Implementation.

UNEP' s strategy on land use management and soil conservation requires close cooperation with governments, civil society and fellow UN and international organisations to ensure a broadly acceptable and efficient implementation, as well as the necessary additional financial, institutional and human resource support.

References

Agnoletti, M. (ed.). *The Conservation of Cultural Landscapes*. CAB International, Wallingford, Oxon, UK. 2006.

Balachandran, Sarojini, ed., *Encyclopedia of Environmental Information Sources,* Gale Research Inc., Washington, DC, 1993

Barth, S., et al., *Exploring the Theory and Application of Ecosystem Management*, University of Michigan, Ann Arbor, MI, 1994.

Cutter, S.L. & Renwick, W.H. *Exploitation, Conservation, Preservation – A Geographic Perspective on Natural Resource Use*. Third edition. John Wiley & Sons Ltd., New York. 1999.

Pepper, D. *Modern Environmentalism*. Routledge, London. 1996.

5

Globalization and Environmental Protection

Globalisation has ushered in an era of contrasts – one of fast-paced change and persistent problems. It has spurred a growing degree of interdependence among economies and societies through transboundary flows of information, ideas, technologies, goods, services, capital, and people. In so doing, it has challenged the traditional capacity of national governments to regulate and control markets and activities.

The rapid pace of economic integration has led to interlinked world markets and economies, requiring a degree of synchronisation of national policies across a number of issues. One dimension of this coordination concerns the environment – from shared natural resources such as fisheries and biological diversity to the potential for transboundary pollution spillovers across the land, over water, and through the air. We now understand that governance approaches that are bounded by the traditional notion of national territorial sovereignty cannot protect us from global-scale environmental threats. An effective response to these challenges will require fresh thinking, refined strategies, and new mechanisms for international cooperation.

Globalisation can have both positive and negative environmental consequences. But the same forces can exacerbate existing environmental problems and create new ones, as well running down stocks of non-renewable natural resources. Economic integration and trade liberalization can generate new resources that permit investments in environmental protection as well as faster and broader dissemination of pollution control technologies and new policy ideas. Environmental choices can likewise shape the path of globalisation. National regulatory choices may act as barriers to liberalised trade, or they may trigger a convergence towards harmonised international standards. The broad range of recent 'trade and environment' disputes at the World Trade Organization (WTO) – over beef hormones, asbestos regulation, genetically modified food, shrimp fishing, and endangered sea turtles, to name a few – highlights the dynamic complexity of these issues. For policymakers, the core challenge lies in finding an appropriate mix of competition and cooperation, market forces and intervention, and economic growth and environmental protection.

To maximize globalization's upside potential, a fundamental reform of global governance structures in general and of the international architecture for *environmental* cooperation in particular will be required. Building greater environmental sensitivity into multilateral trade and financial institutions is necessary but insufficient. An equally broad-scale reform of the global environmental governance architecture is needed. We propose the creation of a Global Environmental Mechanism (GEM) to facilitate efforts to manage global-scale environmental risks; support bargaining and negotiation; promote sound management of the global commons; and advance dissemination of information, 'best practices' in policy-making, and new technologies.

Effects of Globalisation on The Environment

Globalization presents a mixed blessing for the environment. It creates economic opportunities but also gives rise to new problems and tensions. By increasing the volume and decreasing the cost of information, data, and communications, globalization also offers expanded access to knowledge, new mechanisms for participation in policymaking, and the promise of more refined and effective modes of governance. Understanding this array of effects – economic, regulatory, information, and pluralization – is essential if one is to make sense of globalization's impact on the environment.

Economic Effects

Environmental impacts of expanded economic growth and trade can be understood in terms of scale, income, technique, and composition effects. Scale effects refer to increased pollution and natural resource depletion due to increased economic activity and greater consumption. Income or wealth effects appear when greater financial capacity results in greater investment in environmental protection and new demands for attention to environmental quality. With higher income, we observe two other, related phenomena – technique and composition effects. Technique effects arise from tendencies towards cleaner production processes as wealth increases and, as trade intensifies, better access to new technologies and environmental best practices. Composition effects take place as the economic base evolves towards a less-pollution intensive high-tech and services-based set of activities. The overall environmental impact of economic growth depends on the net impact of these four effects. If the income, technique, and composition effects overwhelm the negative scale effect of expanded economic activity, then the impact of growth will ultimately be positive. But in the early stages of industrialization, it may well be that environmental conditions deteriorate.

The precise shape, duration, and applicability of the resulting inverted U-shaped environmental Kuznets curve has generated considerable debate. Grossman and Krueger found the critical income level at which pollution begins to diminish to be about $5000/year per capita GDP. In trying to separate out the various environmental effects of economic growth, Antweiler, Copeland, and Taylor found that, a 1 per cent increase in the scale of economic activity raises pollution concentrations by 0.25 to 0.50 per cent, but the accompanying increase in income drives concentrations down by 1.25-1.50 per cent via a technique effect resulting in improved conditions overall. It appears, however, that expanded trade and economic activity may worsen environmental conditions in other circumstances. Carbon dioxide emissions do not, for instance, appear to fall at any known income level.

Economic theory suggests that the free market can be expected to produce an efficient and welfare-enhancing level of resource use, production, consumption, and environmental protection if the prices of resources, goods, and services capture all of the social costs and benefits of their use. However, when private costs – which are the basis for market decisions – fail to include social costs, market failures occur, resulting in allocative inefficiency in the form of suboptimal resource use and pollution levels. Market failures are a hallmark of the environmental domain. Many critical resources such as water, timber, oil, fish, and coal tend to be underpriced. Ecosystem services such as flood prevention, water retention, carbon sequestration, and oxygen provision often go entirely unpriced. Because underpriced and unpriced resources are overexploited, economic actors are often able to ignore part or all of the environmental costs they generate. Globalisation may magnify the problem of mispriced resources and the consequent environmental harms.

Regulatory Effects

A primary goal of trade liberalisation is the reduction of barriers to market access. Thus, trade agreements often include "disciplines" on how the parties will regulate. Some environmental advocates decry this loss of regulatory sovereignty.

Perhaps more importantly, freer trader promotes competition. Increased competitive pressure may manifest itself in industry or governmental efforts to reduce pollution control compliance costs. This political dynamic could trigger a regulatory 'race to the bottom' in which jurisdictions with high environmental standards relax their regulatory regimes to avoid burdening their industries with pollution-control costs higher than those of competitors operating in low-standard jurisdictions. While there is little evidence that environmental standards are actually declining, the concern is not literally about a race to the bottom, but rather about a race toward the bottom that translates into suboptimal environmental standards, at least in some jurisdictions. Ample evidence exists to support the existence of a regulatory dynamic in which standards are set strategically with an eye on the pollution-control burdens in competing jurisdictions. The outcome may well be a 'political drag' which results in weaker environmental laws than might have otherwise been adopted and, perhaps more importantly, lax enforcement of existing rules or standards.

But, diverse national circumstances generally make uniform standards less attractive than standards tailored to local conditions and preferences. But not always. Divergent standards across jurisdictions may impose market access barriers on traded goods that exceed any benefits obtained by allowing each jurisdiction to maintain individualized requirements. In some cases, producers vying for access to high-standard jurisdiction will drive upward harmonization (a 'race to the top'). But this logic applies only to product standards. Standards for production processes or methods (PPMs) are not subject to the same market pressures.

In an interdependent world, production-related externalities cannot be overlooked. Semiconductors produced using chlorofluorocarbons (CFCs), which contribute to the destruction of the ozone layer, should be treated as contraband. Where international environmental agreements are in place, as with the Montreal Protocol regulating the use of ozone-depleting substances, a recognized standard is available. In such cases, trade rules should be interpreted to reinforce the agreed-upon standards. Recrafted trade principles and World Trade Organization (WTO) rules that accept the legitimacy of environmental controls aimed at transboundary externalities would make global-scale trade and environmental policies more mutually reinforcing and reduce the risk of the trade regime providing cover for those shirking their share of global environmental responsibilities.

Information Effects

One of the key features of globalization is the expansion of communication networks across the globe. The increasing speed and decreasing cost of communication has virtually eliminated the traditional concept of distance. The Information Age has thus transformed space and time, drawing the world into networks of global communication – though some parts are more tightly linked than others. This communication revolution has dramatically increased the intensity of national interdependence, fomenting a greater sense of international community and a foundation of shared values. In turn, the incipient sense of a world community provides citizens with a basis for demanding that those with whom they trade meet certain baseline moral standards, including a commitment to environmental stewardship. As economic integration broadens and deepens, and information about one's partners becomes more readily available, what citizens feel should be encompassed within the set of baseline standards tends to grow.

Increased access to data and information on economic and environmental performance allows for faster problem identification, better issue analysis, and quicker trend spotting. It can also aid the identification of leaders and laggards in the international arena relative to various environmental or social criteria and spur competition (and thus improved performance) among nations, companies, or even communities. Information in and of itself is not, however, necessarily beneficial. Information overload could lead to a cacophony of voices in the policy realm and result in paralysis instead of action.

Pluralization Effects

Intensified interaction in the economic and political spheres coupled with rapidly diminishing costs of communication has increased the number and diversity of participants in global networks. This 'pluralization' is evident in the exponential growth non-governmental organizations (NGOs), their heightened levels of activity, and their increased access to the policy-making process at both the national and international levels. In 1990, there were about 6,000 international NGOs. By the year 2000, that number had reached 26,000. An elaborate organizational or institutional infrastructure is no longer necessary for an entity to have a global reach.

With global interconnectedness on the rise, transparency, participation, and democratization have also increased, providing a broader constituency of concerned groups and individuals with access to global decision-makers. While national governments remain central to global-scale policy-making, many new actors now play a role and the governance process has become much more complex. For example, the landmine treaty resulted from an internet-based campaign started in 1991 by several NGOs and individuals. Today, the treaty has been ratified or acceded to by 141 countries. An NGO network, representing over 1,100 groups in over 60 countries, are now working locally, nationally, regionally, and internationally to implement the ban on antipersonnel landmines. The significance of NGOs as actors on the global stage was recognized in 1997, when the International Campaign to Ban Landmines and its coordinator, Jody Williams, received the Nobel Peace Prize.

The downside of pluralization is that ability to participate in the policy process remains asymmetrical. Constituencies start out with unequal resources and the influence of special interests – which are often well financed and organized – may be magnified. Globalization by no means implies the end of politics. Quite to the contrary, power relations remain important and mechanisms for leveling the playing field become increasingly necessary.

Environmental Effects on Globalisation

Just as globalization will shape environmental protection efforts, so may environmental choices affect the course of globalization, particularly efforts to liberalize trade and investment flows. At one extreme, a rigid harmonization of policy approaches and regulatory standards could run roughshod over diverse environmental circumstances, resource endowments, and public preferences. At the other extreme, uncoordinated national environmental policies might become non-tariff barriers to trade that obstruct efforts to open markets. In these ways, national-level environmental policies may influence international action. Similarly, ecological realities may require policy coordination and collective action on the global scale.

National Activities with International Effects

National environmental performance may have international impacts. In an increasingly interconnected world, environmental harms, such as greenhouse gas emissions, left unattended at the local and national levels may result in global-scale problems, such as the global warming sea level rise, increased intensity of wind storms, and changed rainfall patterns that may come to pass as a result of climate change. The failure to address such spillover of harm creates a risk for the international economic system of being weighed down by market failures. Transboundary pollution spillovers, which result in 'super externalities', are especially difficult to manage. The need to bring multiple countries together in a common response represents a much more difficult problem to address than national-scale environmental protection. As with any global public good, where costs are borne locally and benefits spread across the world, no single jurisdiction has an incentive to regulate such harms optimally. In the case of regular externalities (i.e., harms within one nation), there are many reasons why governments may not optimally regulate emissions or other harmful practices, but at least they have an incentive to do so in the face of the welfare losses of their own citizens. When harms span multiple jurisdictions or even the entire world, there is an increasing likelihood that the government whose facility is causing the negative impact will choose not to act because its own cost-benefit calculus does not justify intervention.

Tensions are also likely to occur when national-scale regulatory policies differ widely among countries that are closely integrated economically. Deeper economic integration makes countries more sensitive to the regulatory choices and social policies of their trade partners. For instance, in the 1970s, when China's trade with the United States (US) totaled less than US$1 billion a year, few US citizens had reason to care about China's labor or environmental policies. Today, as China emerges as a major trade partner and competitor – and US-China trade had increased almost 100-fold to US$92 billion in 2002 – these policies are subject to much greater American interest and concern. A key focus of trade policymaking thus centers on non-tariff barriers to trade and the need for a 'level' playing field in the global marketplace.

Because many domestic regulations could act as non-tariff trade barriers, trade agreements now routinely include market-access rules and disciplines that create a framework for national regulation. Public health standards, food safety requirements, emissions limits, labeling policies, and waste management and disposal rules – all national measures – may shape the flow of international trade. For example, the EU import ban on genetically modified foods has led to a 55 per cent decrease in US corn exports to Europe over the past five years and strenuous US objections to the EU treatment of GMO food. Similarly, Venezuela objected to the discriminatory approach of the reformulated gasoline provisions of the US Clean Air Act of 1990 and won a WTO dispute settlement case restoring its access to the US gasoline market. From the 'Tuna/Dolphin' case of the early 1990s to the recent 'Shrimp/Turtle' dispute, the number of trade-environment flash points has continued to grow. As noted earlier,

In the 'Reformulated Gasoline' case, Venezuela and Brazil brought a complaint against the US alleging that the 'Gasoline Rule', promulgated by the Environmental Protection Agency (EPA) under the Clean Air Act, which excluded importers from exercising two alternatives for determining the appropriate fuel content that were available to domestic refiners, violated the General Agreement on Tariffs and Trade (GATT) as an unjustifiable barrier to trade. In 1996, the Appellate Body of the WTO determined that the 'reformulated' gasoline rule did violate GATT as it subjected Venezuelan and Brazilian refiners to potentially more stringent requirements for fuel emissions than domestic refiners and was, therefore, in violation of Article XX exceptions. Following the decision, the countries agreed on a 15-month phase-out of the illegal regulation.

In the 'Tuna/Dolphin' case, US import restrictions on tuna caught with unsafe nets and techniques were struck down under the GATT rules as an illegal barrier to trade. Under the Marine Mammals Protection Act of 1972, the US restricted the importation of tuna caught using methods that killed dolphins. The restrictions effectively imposed a barrier to trade on tuna caught in Mexico as a result of the ban on such importation. Mexico successfully argued that the ban served as an illegal barrier to trade under GATT and that the US could not extraterritorially regulate in the name of the environment.

In 1996, the US Court of International Trade ordered the prohibition of shrimp importation from all countries that had not adopted harvesting methods comparable to the US methods, which included Turtle Exclusion Devices to prevent further mortality of endangered sea turtles. India, Malaysia, Pakistan, and Thailand brought issue with these Guidelines at the WTO. In 2001, upon Appellate review, the WTO issued the ruling in the 'Shrimp/Turtle' case, upholding, for the first time in GATT history, unilateral trade restrictions to conserve extraterritorial natural resources. The restrictions were upheld under the General Exceptions in GATT Article XX. The outcome is contrary to that in the 'Tuna/Dolphin' dispute as sea turtles had been listed by the United Nations as threatened with extinction.

Environmentalists fear that liberalized trade might make it harder for high-standard countries to keep their stringent environmental requirements in the face of market-access demands from trade partners.

The essential difficulty lies in separating legitimate environmental standards from protectionist regulations advanced under the guise of environmental protection. Few would argue, for example, that emission-control standards for cars are an unwarranted barrier to trade. However, the fear of protectionism in an environmental disguise is not unfounded and needs to be addressed, particularly if developing countries are to retain confidence in the fairness of the international trade system. The smooth functioning and efficiency of the international economic system cannot be maintained unless there are clear rules of engagement for international commerce, including environmental provisions.

Global Environmental Policies

Globalization is, in part, an ecological fact. There exist a series of environmental challenges that span multiple countries and even the globe. Polluted waters, collapsing fisheries, invasive species, and the threat of climate change are all realities that have been exacerbated by globalization. But, ecological realities also affect the pace and pattern of globalization. Scarce environmental resources, such as water, shape countries' perceptions of their independence or interdependence and consequently influence their economic and political interactions within the global community. The value that citizens around the world place on nature and biodiversity within foreign jurisdictions may spur international political pressures that limit a country's economic and regulatory choices. Protection of the shared resources of the global commons – the oceans, the atmosphere, etc. – provides a rallying point for NGOs aiming to promote worldwide collective action. Increased understanding of the interdependence of ecological systems contributes to establish a more robust global environmental regime.

Clearly, the primary responsibility for environmental protection rests with national governments and local communities. But some problems are inescapably regional or global in scope and cannot be addressed without international cooperation. Yet, incentives to pursue behavior that is individually rational but collectively suboptimal are especially strong with regard to the depletion of natural resources, which at once may be seen as belonging to everybody and nobody. It is rational for a fisherman, for example, to try to maximize his personal gain by catching as many fish as possible as quickly as possible.

Collectively, however, such a strategy leads to overexploitation of the resource and a 'tragedy of the commons,' leaving the entire fishing community worse off than if it had found a cooperative arrangement to manage the fishery on a sustainable basis. When extended to a global scale, the problem becomes even more acute and intractable in the absence of clear rules and institutions ensuring sustainable resource management. Such global-scale issues require responses aggregated beyond the level of national jurisdictions or, at the very least, coordinated national action.

While not strictly necessary, international cooperation is helpful in attacking a set of common problems encountered locally all across the globe and thus of concern to policymakers the world over. These problems – including control of air and water pollution, no waste disposal, etc. – should be dealt with by local or national authorities. There is inherent need for global-scale cooperation.

But the fact that many countries face problem in common creates another logic for cooperation – the potential to gain from sharing data, information, and policy experiences. Comparative analysis often helps to illuminate issues and highlight best practices – policies and technologies – to be deployed in response. To the extent that a problem requires substantial scientific or technical analysis, cooperation may also generate economies of scale in data collection, analysis, and other research functions both benefiting from globalization and contributing to a deepening of interconnectedness and interdependence.

Global Environmental Governance

Without effective international-scale governance, globalization may intensify environmental harms wherever national regulatory structures are inadequate. In strengthening competitive pressures across national borders, economic integration may help consumers by lowering prices, improving service, and increasing choice. But these same pressures at times threaten to overwhelm the regulatory capacities of national governments and thus necessitate intergovernmental coordination of domestic policies and cooperative management of the global commons. As shown above, some problems are local and can best be addressed on that scale But even in these cases, there is a clear advantage of learning from other countries and localities that have managed to address similar issues. In other cases, the problems are so inextricably international that a coordinated multi-country response is required. This response, however, must always be backed up by effective action at the national and local levels.

Theory suggests that the solution to this policy dilemma lies in a structured program of collective action. But overcoming the collective-action problem is especially difficult in the international realm. There is no Leviathan or overarching authority. And while the number of beneficiaries and potential contributors to a global public good may be much larger than on the national scale, so too is the number of potential contributors to a public 'bad'. The spatial and temporal distribution of causes and effects makes it hard to identify those who fail to cooperate. Moreover, in the absence of an international authority, even if defectors were detected, there are scant means of discipline and sanction. The problem, therefore, is one of organizing and maintaining cooperation. Absent institutional support and efforts at collective action tend to degrade towards what is called in game theory a lose-lose or Nash equilibrium. The situation must be converted from one in which decisions are made independently based on narrow self-interest to one in which actors adopt cooperative solutions serving a broader, common interest.

The traditional policy prescriptions – a set of taxes or subsidies to internalize externalities – cannot be easily applied to a multi-jurisdictional context with a fragmented institutional structure. Successful intervention requires some mechanism for promoting collective action. Fragmentation, gaps in issue coverage, and even contradictions among different treaties, organizations, and agencies with competing responsibilities have undermined effective, results-oriented action in the domain. As pointed out by Charnovitz, '[l]ike a city that does not have zoning ordinances, environmental governance spreads out in unplanned, incongruent, and inefficient ways.' A pervasive lack of data, information, and policy transparency adds to the challenge. An institutional structure is necessary that can provide: the data foundation needed for good environmental decision-making; the capacity to gauge risks, costs, benefits, and policy options comparatively; a mechanism to exert leverage on private-sector and governmental resources deployed at the international level; and means to improve results from global-scale environmental spending and programs.

Reforms in Environmental and Economic Governance

While the UN Environment Program (UNEP) lies at the centre of the environmental regime, international environmental governance falls within the mandate of multiple organizations in the United Nations (UN) system. Hampered by a difficult mandate, a modest budget, and limited political support, UNEP competes with more than a dozen other UN bodies, including the Commission on Sustainable Development, the UN Development Program (UNDP), the World Meteorological Organization (WMO), and the International Oceanographic Commission on the international environmental scene. Adding to this fragmentation are the independent secretariats to numerous conventions, including the Montreal Protocol (ozone-layer protection), the Basel Convention (hazardous-waste trade), the Convention on International Trade in Endangered Species, and the Climate Change Convention, all contending for limited governmental time, attention, and resources.

The existing international environmental system has failed to deal adequately with the priorities of both developed and developing countries. The proliferation of multilateral environmental agreements has placed an increasing burden of collective obligations and responsibilities on member states. The toll on developing countries has been especially heavy as little assistance in the form of financing, technology, or policy guidance has been forthcoming. The inadequacy and dispersion of the existing financial mechanisms – scattered across the Global Environment Facility, UNDP, World Bank, and separate funds such as the Montreal Protocol Finance Mechanism – reinforces the perception of a lack of seriousness in the North about the plight of the South. Furthermore, fundamental principles of good governance such as participation, transparency, and accountability are still at issue in many of the institutions with environmental responsibilities. These procedural shortcomings undermine the legitimacy of the system as a whole.

In the absence of a functioning global environmental management system capable of addressing the growing number of international environmental issues, environmental groups have directed efforts towards the reform of international economic bodies, including the World Bank and the WTO. The WTO has been of particular interest as it has assumed responsibility for integrating the policy realms of environment and trade. Although the WTO has a Committee on Trade and Environment that has been meeting for a number of years, the WTO dialogue has been dominated by trade experts, has demonstrated little understanding of the impact of trade on environmental policy, and has almost nothing in the way of results to show for its efforts. The role of the WTO as the principal forum for the discussion and resolution of trade and environment concerns has been contested by both the environmental community and developing countries. Environmentalists perceive the WTO as an organization charged narrowly with the promotion of trade liberalization and argue that any attempt to mainstream environmental issues within the WTO inevitably privileges economic concerns over the environment. Free traders, on the other hand, regard the WTO as an inappropriate forum for environmental issues, which they see as burdening the trade regime. Developing countries, too, see the inclusion of environmental rules among the responsibilities of the WTO as a complication and a threat, potentially creating an excuse for protectionism and the exclusion of Southern goods from Northern markets. Nevertheless, discussion is taking place within the WTO, and pressure to 'green' the organization has resulted in a number of notable reforms.

Recognition of the WTO's lack of capacity for addressing environmental issues and the undermining of its efficacy and legitimacy whenever the organization is forced to make decisions that go beyond the scope of its trade mandate and expertise have led a number of trade experts to call for the creation of a more robust environmental governance structure. The former WTO Director-General, Renato Ruggiero, and the current Director-General, Supachai Panitchpakdi, have both urged for the creation of a World Environment Organization to help focus and coordinate worldwide environmental efforts. During the World Summit on Sustainable Development in 2002, French President Jacques Chirac called for the creation of a Global Environmental Organization that would bring greater balance to a multilateral system excessively focused on the economy. Similar calls have come from Mikhail Gorbachev, Lionel Jospin, The Economist magazine, and others. It is becoming increasingly clear that successful reform of the trade and finance system needs to be coupled with an equally rigorous and fundamental reform of the global environmental regime.

Governance Alternatives

Collective action in response to global environmental challenges continues to fall short of public needs and expectations as a result of the deep-seated weakness of the existing institutional architecture. The question, therefore, is not whether to revitalize the global environmental regime, but how. The integrated and interdependent nature of the current set of environmental challenges contrasts sharply with the nature of the institutions we rely upon for solutions. These institutions tend to be fragmented and poorly coordinated, with limited mandates and impenetrable decision-making processes.

Shifting from a prisoners'-dilemma world of free-riding and lose-lose outcomes to one where reciprocity is recognized and collaboration understood will require careful institutional realignment. We need an approach that acknowledges the diversity and dynamism of pollution control and natural-resource-management problems and recognizes the need for specialized responses. The multi-faceted nature of the environmental challenge requires a multi-layered institutional structure that can address issues on various geographic scales and with a variety of policy tools.

Functions at Various Levels of Governance

We argue that there is a spectrum of global-governance responses ranging from very light to fairly robust. Amenable to a regime at the light end of the spectrum lie problems that are local in scope but can be found around the world (local water and air pollution, for example). As we move towards the more demanding side of the spectrum, regional issues such as international water-bodies pollution or regional fisheries management arise. At the most difficult end of the spectrum are issues that likely require a strong structure of global collaboration (climate change, ozone layer, ocean pollution). A number of functions need therefore to be performed at the various levels of governance by different institutions.

When dealing with global-scale problems, institutions need to possess several capacities, including the ability to identify and define problems, raise awareness about them in various forums, draft rules and create norms for behavior leading to the solution of these problems, formulate policy options, facilitate cooperative actions among governments and other actors, finance and support activities, and develop management systems. As will be elaborated below, we see an information clearinghouse, a technology clearinghouse, and a policy forum as central elements to the effective functioning of a global regime for resolution of environmental problems. Global institutions also have an important role to play when the problems are primarily national in scale. They can serve as facilitators of information and knowledge exchange, promoting learning across contexts and among actors. The exchange of data, best practices, policies, and approaches could be an important tool in problem solving at the national level.

National institutions also have roles to play, both at domestic and global levels of governance. National governments remain the primary actors charged with regulatory and enforcement powers to solve environmental problems. Functions such as standard setting, policy formulation, compliance monitoring, and evaluation are among their responsibilities. When the problems are of a global character, national governments are again key actors. Implementation of multilateral agreements is ultimately their responsibility. They also engage in information-sharing and exchange in the process of arriving at agreement on the global problems to be addressed, the policies necessary for their resolution, and the actions to be undertaken domestically.

An effective response to both the common elements of national problems and the special demands of transboundary issues requires a deft and agile structure able to hone in on the nature of problem and produce the right scale of activity while promoting worldwide cooperation. There is no silver bullet. Various institutional and organizational designs are possible. We believe that the best strategy centers on a new environmental mechanism at the global level. Conceptually, a global environmental mechanism (GEM) would fundamentally need to focus on promoting collective action on the international scale. Practically, it offers the chance to build a coherent and integrated environmental policy-making and management framework that addresses the challenges of a shared global ecosystem.

We see three core capacities as essential: (1) provision of adequate data and information that can help to characterize the problems to be addressed, reveal preferences, and clarify reciprocity; (2) creation of a policy 'space' for environmental negotiation and bargaining; and (3) sustained support for national efforts to address issues of concern and significance. We identify data collection, monitoring, and scientific assessment as central in the information domain. A forum for issue linkages and bargaining, a mechanism for rule-making, and a dispute-settlement framework are essential to ensuring cooperative solutions. The continual development of technical, financial, human, and institutional capacities for addressing diverse challenges is another critical function requiring effective institutional mechanisms at the global level.

At present, various institutions and agencies ostensibly have many of the identified capacities. But the reality often falls short of the promise. And some are flagrantly absent. For example, a host of international organizations, scientific research centers, national governments, and environmental convention secretariats are carrying out data collection, scientific assessment, financing, and technology transfer with little coordination across jurisdictions. Compliance-monitoring and -reporting are unsystematic, scattered, and largely informal. The participation of non-state actors requires further structural elaboration and institutionalization along with procedures for rule-making. A forum for issue linkage, bargaining, and trade-offs as well as a dispute-settlement mechanism is lacking. A more robust policy space for the environment is necessary to sustain efforts at environmental advocacy within the broader system of global governance and to ensure that environmental concerns are integrated into sustainable development policies.

Building on the expertise and capacities of existing institutions and creating new mechanisms where functions are not currently performed, we see three institutional elements as central to a successful global environmental system. A Global Information Clearinghouse might represent a first step towards improved global environmental governance through provision of comparable data on environmental quality, trends, and risks. The coordination of existing institutional mechanisms for data collection, scientific assessment, and analysis might attract broad-based support. A Global Technology Clearinghouse, focusing on information-sharing, performance measurement and benchmarking, and dissemination of best practices, might also be launched as an early initiative with likely broad appeal. With competence established in these areas, a Global Bargaining Forum might be initiated with the capacity for rule-making and facilitation of burden-sharing. Progressive development over time, as the new system proves its capacity and value, is likely to make any governance-reform strategy more acceptable to nations reluctant to yield responsibility or control to any global entity.

Global Environmental Information Clearing-House

Better environmental data and information make it easier to identify problems and trends, evaluate risks, set priorities, establish policy options, test solutions, and encourage technology development. A global information clearinghouse providing timely, relevant, and reliable data on environmental issues and trends could transform the policy-making process on the global scale. Better data, science, and analysis could shift assumptions, highlight preferences, and sharpen policies. In the case of acid rain in Europe, for example, knowledge of domestic acidification damage allowed for refined policies that triggered emission reductions in several countries. Simply put, data can make the invisible visible, the intangible tangible, and the complex manageable.

Information on how others are doing in reducing pollution and improving resource productivity tends to stimulate competition and innovation. Comparative performance analysis across countries – similar to the national PROPER scheme in Indonesia– could provide much greater transparency, reward policy leaders, and expose laggards. Just as knowledge that a competitor in the market place has higher profits drives executives to redouble their efforts, evidence that others are outperforming one's country on environmental criteria can sharpen the focus on opportunities for improved performance. The attention that the Yale-CIESIN-World Economic Forum Environmental Sustainability Index has generated demonstrates this potential.

Data-gathering should primarily be the function of local or national organizations. But a central repository for such information and a mechanism for making the information publicly available could generate significant economies of scale, efficiently generate relevant comparisons, and expose slack performance. An information clearinghouse would not centralize science policy functions but create a centralized source for coordinating information flows among the institutions responsible for performing scientific aspects of policy-making.

Technological Advances and Environmental Conservation

Globalization is fuelled by and plays a central role in the diffusion of technologies. Technological advances are often the key to environmental gains. However, industrialized countries dominate the technology market and the generation of innovations. Some technologies and their environmental features may, therefore, be inappropriate for the economic and environmental circumstances of less developed countries.

Most multilateral environmental agreements contain provisions related to technology transfer as part of the incentive packages for developing countries to meet their obligations under the conventions. The Basel Convention on the Control of Transboundary Movements of Hazardous Wastes and their Disposal, the Montreal Protocol on the Ozone Layer, the Convention on Biological Diversity, the Framework Convention on Climate Change and its related Kyoto Protocol all cite technology transfer as a critical method for achieving concrete environmental improvements. Agenda 21 also underscores the importance of technology transfer to sustainable development. The existing strategies for technology transfer have, however, been less than effective. A new mechanism to bring technologies to developing countries must be part of any strategy to improve international environmental policy results. Establishing such a mechanism, however, presents a significant challenge.

The empirical evidence shows that the gains from such cooperative arrangements have indeed been significant and beneficial for the environment. For example, the technology panel convened under the Montreal Protocol to report on the availability of CFC substitutes and the feasibility of larger production cuts generated new knowledge and new commercial opportunities for CFC reduction in a highly collaborative process. Most technologies are, however, owned by private companies not governments. So some effort need to be put into structuring incentives to motivate the private sector to disseminate technological advances optimally. An effective environmental technology clearinghouse is thus not only necessary but also possible. It would contain information on best practices around the world and facilitate technology development and continuous learning.

Responses to Global-scale Environmental Problems

Successful responses to global-scale environmental problems depend on effective international agreements. To be workable, any such agreement must equitably distribute the burden of international collective action. Developing countries will often need support, subsidies, and other incentives to encourage their efforts to internalize externalities. In the past, issue linkage has been avoided in favor of lowest-common-denominator programs in the absence of funding to support those least well positioned to act. As Whalley and Zissimos argue, there would be great value in a forum for the facilitation of international deals on the environment that improve quality and result in positive cash flow to custodians of environmental assets.

A global bargaining forum could act as a catalyst for action, facilitating financial discussion among countries or private entities. A government in one country might, for example, negotiate a deal to preserve a particular natural resource in another country in return for a sum of money or other policy benefits. Brazil might, for instance, commit to certain limits on development in the Amazon in return for guaranteed access to European and US markets for its orange juice. The forum might also provide mechanisms for verification, financial transfers, and dispute settlement.

In designing a new global environmental architecture, form should follow function. We envision a 'light' institutional superstructure providing coordination through a staff comparable in size and quality to the WTO secretariat in Geneva. The secretariat's primary role would be to promote cooperation and achieve synergies across the disparate multilateral environmental agreements and other international institutions with environmental roles. A properly designed structure would provide a counterpart as well as a counterweight to the WTO and an alternative forum for addressing tensions over divergent environmental values and approaches. The GEM we envision would neither add a new layer of international bureaucracy nor create a world government. Quite to the contrary, movement towards a GEM should entail consolidation of the existing panoply of international environmental institutions and a shift towards a more modern 'virtual' environmental regime.

At the centre of our proposal lies a global public-policy network drawing in expertise from around the world on an issue-by-issue basis. By utilizing the resources of national governments, NGOs, private-sector enterprises, business and industry associations, think tanks, research centers, and academic institutions on an 'as needed' basis, the GEM would have far broader issue expertise and analytic capacity than has the existing environmental regime. Such a system for advancing international environmental agenda-setting, analysis, negotiation, policy formulation, implementation, and institutional learning would be more flexible, cost-effective, fleet-footed, and innovative. The benefits of such a structure are increasingly clear. Global public-policy and issue networks respond to an ever more complex international policy environment, taking advantage of Information Age communication technologies to draw in relevant expertise, analyze problems from multiple perspectives, and build new opportunities for cooperation.

Streamlining the environmental system would be especially beneficial to the South. In particular, a single venue for negotiations and international coordination would make it much easier for the overstretched environment ministries of the developing world to monitor the spectrum of environmental issues at play and to contribute thoughtfully to the global-scale debate even with a relatively small international policy-making team. There would be no need to traipse around the world trying to keep up with an ever more extensive list of separate bodies and meetings. A network approach, drawing in diverse perspectives and expertise and using the Internet, could facilitate greater developing country participation in the international policy-making process.

Who will pay for global-scale environmental problem-solving stands out as a matter of particular importance to developing countries. Globalization, as noted above, puts increasing pressure on national governments to become more competitive in the global marketplace. Expending scarce financial resources for environmental protection is, therefore, often regarded as counterproductive by developing countries, especially if there is no urgent demand from domestic constituencies. By placing the principle of common but differentiated responsibilities at the centre of the new mechanism along with a real forum for bargaining and trade-offs, efforts to strike a fair balance of rights and responsibilities with regard to transboundary environmental issues might meet with increased success. A more carefully considered and coherent set of international environmental standards would also alleviate fears in the South that the industrialized world seeks to impose unreasonably high standards – and perhaps trade penalties for non-compliance – on developing countries, all of whom have many competing demands for limited public resources. Moreover, mechanisms to support technology transfers and to subsidize developing countries' environmental initiatives in pursuit of global environmental goals would help to alleviate North-South tensions.

A related question concerns the values to be promoted in a strengthened international environmental regime. It is essential that a GEM be seen as a transparent and inclusive forum that seeks to build consensus on a basis that respects the diversity of views across the world. Properly managed public policy networks create 'virtual public space' that is easier to enter than the established physical fora where decisions are currently made. An Information Age set of outreach mechanisms could also decrease the distance between decentralized constituencies and global decision-makers – making it easier to insert into the policy process the broad array of values, perceptions, and perspectives that are now often overlooked or incompletely considered. At the same time, these mechanisms would facilitate public understanding of the issues addressed and decisions made on the global scale.

Both economic and ecological interdependence require rigorous national policies and effective international collective action. Our increasingly globalized world makes new thinking about international environmental cooperation essential, both in its own right and to undergird further economic integration. An extraordinary mix of political idealism and pragmatism will be required to coordinate pollution control and natural-resource-management policies on a worldwide basis across diverse countries and peoples, political perspectives and traditions, levels of wealth and development, beliefs and priorities. But the gains to be achieved go beyond the environmental domain. Indeed, coordinated.

References

Runge, F. 'A global environmental organization (GEO) and the World Trading System,' *Journal of World Trade,* 35(4), 399-426. 2001.

Speth, J. (ed.), *Worlds Apart: Globalization and the Environment,* Washington, DC: Island Press. 2003.

Witte, J., C. Streck, and T. Benner (eds), *Progress or Peril? Partnerships and Networks in Global Environmental Governance.* The Post-Johannesburg Agenda, Washington, DC/Berlin: Global Public Policy Institute.2003.

Wofford, C., 'A greener future at the WTO: the refinement of WTO jurisprudence on environmental exceptions to the GATT', Harvard Environmental Law Review, 24: 563. 2000.

6

Forest Land Conservation

In 1992, the Earth Summit in Rio de Janeiro highlighted the environmental destruction occurring in the world and the need to protect biodiversity hotspots. Following on the tail of the Bruntland Commission report, the Earth Summit called for sustainable development and greater protection of valuable ecosystems. Set in Brazil, the Earth Summit also drew attention to the plight of the world's forests as images of the Amazon aflame awakened many to the threats of rampant deforestation. As a result, 150 government leaders signed the Convention on Biological Diversity (CBD) that emerged as a major outcome of the Rio meeting. As part of the CBD, the participating governments agreed to "[e]stablish a system of protected areas or areas where special measures need to be taken to conserve biological diversity." In the same year, participants in the Fourth World Congress on National Parks and Protected Areas agreed to designate a minimum of 10% of each biome under their jurisdiction (oceans, forests, tundra, wetlands, grasslands) as protected areas.

Establishment of Protected Areas

The establishment of protected areas to protect forest lands is largely based on a belief that government jurisdiction over forests with defined restricted uses is necessary for sustained conservation. The desire to protect large territories so habitats can be connected at a large spatial scale is also a scientific reason often given as a foundation for the creation of protected areas. Many of the larger reserves include a multiplicity of governance types with international donor funding used to initiate planning at a regional scale. While there is validity to the argument that preserving habitat for some species should be done at a large spatial scale, the preference articulated by many conservationists for government protection does not have as strong a foundation.

Managing protected areas has been seen by many as the preeminent method for protecting forests, wildlife, and wilderness in general. Today, more than 100,000 protected areas have been launched and officially encompass roughly 10% of the world's forests. During the last half century, developing countries greatly expanded the extent of their land designated as protected areas. In many instances, however, the designation of a protected area on a map generated substantial donor funding, but not the creation of effective protected areas on the ground. Donor projects have tended "to invest heavily in extensive background studies and elaborate plans made by outside experts, generating reports that too often are rarely used, culturally irrelevant, and quickly obsolete."

Some conservationists continue to call for an increase in the area of forests under strict reserve management. However, we actually do not know how well protected areas conserve the lands, so how can we in good conscience extend this system to the rest of the world's forests? Considerable debate exists regarding the general causes of loss of biodiversity or the massive levels of deforestation that are occurring. Many cases of government weaknesses in managing natural resources have been documented.

In 1999, the World Conservation Union reported on the effectiveness of forest protected areas and concluded that protected areas continue to face threats from human pressures and legal designation does not ensure sustained conservation. The survey that the International Union for Conservation of Nature and Natural Resources (IUCN) conducted of protected areas in 10 key forested countries found only 1% of these protected areas were secure from threat. The IUCN further noted that many protected areas lack financial and human resources, a supportive legal framework, and the institutional infrastructure necessary to regulate agriculture, grazing, forestry, mining, hunting, civil conflict, and tourism.

A recent study conducted by WWF International of more than 200 protected areas in 37 countries found similar results. While protected areas in these countries are legally well established and many of the boundaries are demarcated, these formal arrangements are not sufficient to protect an area by themselves. Among the weaknesses identified by survey respondents was the effectiveness of their own protection systems. Thus, they frequently failed to monitor and enforce the reserve regulations. Protected areas also consistently failed to engage in positive relations with the local residents and with indigenous peoples. WWF found that four key threats endanger forest protected areas: poaching, encroachment, logging, and gathering of non-timber forest products.

The poor results for protected areas in Latin America are environmentally important, because 60% of the world's tropical forests lie in this region. Furthermore, according to Michael Jenkins, president of Forest Trends, and his colleagues, approximately 90% of the world's forests remain outside protected area systems. Should the first priority be to place these forests within protected areas? Or, are there other options that should be considered? Due to the many problems encountered by protected areas that exclude local residents from any formal role in relationship to the designated areas, many analysts, donors, and environmental groups have successfully urged the creation of integrated conservation and development projects that have dual responsibilities: to protect biodiversity and to enhance the economic development of people living around a protected area. While some of these projects have succeeded in achieving aspects of both goals, many have been based on unrealistic assumptions.

Political and Economic Costs of Protected Areas

As both the IUCN and WWF reports conclude, protected areas struggle to monitor and enforce forest regulations adequately. In addition, one of the greatest challenges for protected area personnel is working with local communities to mediate human pressures on ecological resources. Managing protected areas is particularly difficult when active local resistance to protected area policies is present and the protected areas have limited financial and human resources.

Unfortunately, through much of history, decisions to create government-protected areas have often been made by conservationists or colonial powers with little thought about the rights, cultural traditions, and livelihood needs of the local residents living in the ecological regions. Today, many advocates continue to promote the use of protected areas irrespective of local livelihood needs, declaring that biodiversity protection is a moral imperative that can only be adequately protected through strict government regulations and that sustainable development and ecologically friendly communities are mere myths—at least in the naïve way that many integrated conservation and development programs have been funded.

Inattention to the political and economic costs of protected areas, however, leads some advocates to believe that simply declaring a territory to be a protected area is sufficient for all conservation needs and ignores the challenges many of these areas face. The costs of enforcing laws that are not perceived to be legitimate by those expected to comply with the laws has repeatedly been found to be excessive. The problem of hiring guards or police, paying them well, imposing costly sanctions on those caught breaking the law, trying to ensure that guards do not use opportunities to collect bribes, and coping with widespread dissatisfaction with what is conceived as illegitimate imposition of formal laws is a general problem. It is not restricted only to the protection of forests.

In her study of rates of compliance by taxpayers with government-imposed taxes, for example, Margaret Levi uses the concept of "quasi-voluntary compliance" to explain why in some countries citizens do comply with taxes at very high levels. Paying taxes in these systems is "voluntary" in the sense that many citizens choose to comply in situations where they are not being directly observed and coerced. It is not entirely voluntary, however. It is only *quasi*-voluntary since "the noncompliant are subject to coercion—if they are caught." It is possible to achieve a general strategy of quasi-voluntary compliance where citizens have confidence that "(1) rulers will keep their bargains and (2) the other constituents will keep theirs. Taxpayers are strategic actors who will cooperate only when they can expect others to cooperate as well. The compliance of each depends on the compliance of the others." Understanding these conditions is crucial for those interested in protecting forests and other natural resources.

When protected areas are declared in some distant capital by officials who fail to consider or inform local populations, residents of the reserve may not even know a protected area exists. Furthermore, even when locals are informed of the protected status, those who have relied on the resources for their own livelihoods for long periods of time, or perceive it to be their right to exploit the natural resource system, may continue their old practices and engage in violent protests when officials are sent to enforce a law that is not perceived locally as legitimate and is not consistently enforced.

When residents do not believe that the government has the right to regulate their resource use, they will often find ways to resist or sabotage park regulations. Conflict between park residents and park personnel is well-documented and a consistent theme in discussions about protected areas. Examples of places where conflict exists between residents and park personnel include Khoa Yai in Thailand, where local residents resisted protected area policies implemented by the Royal Forestry Department and fighting resulted in the deaths of local residents and park personnel. In Eastern Africa, the Maasai have protested protected area regulations that eliminated key watering spots and disrupted their traditional cattle herding patterns. Similarly, in Costa Rica, park policies in certain regions have enraged local residents who feel that the park administration is impinging on local livelihoods. And, in the Rio Plátano Biosphere Reserve in Honduras, as in other contested protected areas, park guards frequently receive death threats and active resistance to park regulations. Protected areas in general, and contested protected areas in particular, are economically costly to monitor and enforce. Costs to manage protected areas are increasing as current global trends indicate that public expenditure and international financing are flat or declining. The conservation community estimates that an additional U.S. $27 to $30 billion is needed annually to adequately manage protected areas.

In recent years, more attention has been given to including greater local participation in protected areas in order to reduce conflict, support traditional conservation practices, and decrease monitoring and enforcement costs. This is consistent with a broadening of the concept of property rights themselves. Adrian Phillips, former chair of the World Commission of Protected Areas, stated that the crucial lesson in protected area management is "the iron rule that no protected area can succeed for long in the teeth of local opposition." But, many protected area planners and administrators are still unable to enact participatory policies that are legitimate in the eyes of the residents.

A significant gap in the analysis of forest protection is attention to other institutional mechanisms for conservation. A survey of forest management by Molnar and colleagues finds that a minimum of 370 million hectares of global forest lands are under community conservation. Their work and the work of others demonstrate that public ownership is not the only institutional arrangement that may be associated with environmental conservation. The number of forests lying outside protected areas also demonstrates the need to understand what conditions have promoted their protection and thwarted their destruction. The Forest Trends study concludes that secure tenure rights, institutional and regulatory support for community institutions, fair access to markets, direct finance to local communities, and engagement of local communities in conservation research all appear to increase the probability of successful community forest conservation. The findings suggest the importance of considering conservation prospects outside of legally designated protected areas.

The Need to Consider Alternatives

Despite the apparent importance of local communities in forest conservation, many conservationists are reluctant to step outside the confines of the protected area model and explore alternative institutional arrangements for forest management.

For example, in their study published in *Science* on the *Effectiveness of Parks in Protecting Tropical Biodiversity*, Aaron Bruner and colleagues examine the ability of parks to mediate anthropogenic threats. Regrettably, Bruner and colleagues fail to compare parks to these alternatives. Instead they base their findings on a survey of park officials about the conditions inside their own parks and within a ten-kilometer boundary outside the parks. The authors find that protected areas are effective, particularly when parks are actively monitored and enforced by official guards. Relying on park officials alone to judge the effectiveness of their own park is, however, subject to considerable methodological concerns. Based on these questionnaire responses, Bruner and colleagues conclude that central, law-defined, strictly protected area systems enforced by public officials are necessary. Unfortunately, they fail to consider the effectiveness of alternative conservation strategies.

The principal arguments given for protected areas as the only way to conserve forests are that (1) the only way to maintain forest cover is the establishment of defined areas that are owned and regulated by a national government for the purpose of preservation; (2) resource users are unable to create and enforce appropriate resource management rules; and hence (3) substantial investment in top-down enforcement is essential to achieve adequate environmental protection. Results from studies on forest management, however, suggest that these three arguments are myths that do not hold when tested with empirical evidence.

Empirical Studies of Diverse Forest Institutions

To examine these myths, we draw on multiple studies that colleagues associated with the Center for the Study of Institutions, Population, and Environmental Change (CIPEC) and the Workshop in Political Theory and Policy Analysis, both at Indiana University, have conducted as a result of their collaboration with an international network of scholars interested in understanding how institutions interact with biophysical and behavioral factors to influence land use and land-use change—particularly forested land. In these studies, we have measured forest conditions using multiple measures.

International Forestry Resources and Institutions Field Protocols

The International Forestry Resources and Institutions (IFRI) research and training program was initiated as a result of a request in 1992 from Dr. Marilyn Hoskins, who headed the Forestry, Trees, and People Program at the Food and Agricultural Organization (FAO) of the United Nations. Many policies were under discussion at FAO related to the best legal structure to develop and to enhance forest preservation. Dr. Hoskins asked us to develop a reliable methodology that could be implemented by research teams in developing countries working together as a network to explore the effectiveness of protected areas, community forests, and diverse types of government forests by conducting well-designed empirical studies in multiple regions of the world. FAO was aware of our work on irrigation and other resource institutions and hoped that the theoretical foundations of that work could be applied to the study of forestry institutions.

Our research team spent two years developing a series of ten protocols that forests to be studied and good information about how institutions related to forest governance were devised and monitored, and whether or not they were successful in the field. In the process of designing our protocols, we involved the advice and input from more than 100 researchers and policymakers located in all regions of the world.

We consulted officials who were implementing National Environmental Action Plans to ascertain the types of information they needed for future policy. We worked with several forest departments to add measures that were of importance to them. We sought the help of researchers who have a major interest in questions of biodiversity and forest sustainability as well as the impact of institutional arrangements.

The core set of ten protocols that resulted from this wide consultation process is designed to enable scholars to examine the impact of diverse ways of owning and governing forests (such as individual ownership, joint ownership by a community, and different forms of government ownership) on investment, harvesting, protection, and managing activities and their consequences on forest conditions, including biodiversity. We have developed a large relational database that is used to record structured and qualitative data in a consistent manner across sites. The core protocols are designed so that additional questions can be addressed in specific studies designed by collaborating researchers by adding a specific set of questions to one of the existing protocols, or by adding a new protocol such as a household survey.

A long-term collaborative research network has now been established with centers located in Bolivia, Colombia, Guatemala, India, Kenya, Mexico, Nepal, Tanzania, Thailand, Uganda, and our own center in Bloomington, Indiana. Research using the IFRI protocols has also been conducted in Brazil, Mali, and Madagascar. Together, colleagues in the network have collected data in more than 200 sites and revisited 41 of them at least one time. As of December 1, 2004, we have taken a random sample of 8695 forest plots and actually measured 127,712 trees (including diameter, height, and species). This has involved a very large effort to devise appropriate sampling plans, locate the sample plots in forests, and measure the trees, shrubs, and groundcover in the plots with regard to ecological measurements.

An independent forester or ecologist is asked to rank the vegetation density of the study forest in comparison to other forests in the same ecological zone. The vegetation density in the forest under study is ranked on a five-point scale from very sparse to very abundant. Similarly, the forest users are asked to rank forest condition from very sparse to very abundant. These qualitative assessments provide a measure of forest cover that can be used to compare forests across ecological zones. In the studies discussed below, we specify whether the forester's assessment and/or the community assessment is used to determine forest vegetation density.

Analysis of Changes in Forest Cover Reflected in Over-Time Satellite Images

CIPEC researchers also have actively used remotely sensed data in many of our research efforts. The analysis of several remotely sensed images over a decade or more is a particularly useful technique for examining the impact of institutional arrangements, as we illustrate below. In our discussion of Myth 1 below we present a multi-temporal composite of the Maya Biosphere Reserve in northern Guatemala developed by Edwin Castellanos, Glen Green, and Victor H

Challenging of Forest Protection

Findings by CIPEC researchers challenge the beliefs that protected areas are the only way to protect forest lands and that resource users are unable to enforce or create forest management rules. Deforestation is driven by a complex web of factors acting at local, national, and global scales. Demographic, economic, technological, institutional, cultural, and sociopolitical forces all interact with biophysical features of the land to produce patterns of deforestation. Empirical studies of forest management demonstrate that while in certain conditions protected area policies are effective in controlling deforestation, they are not foolproof solutions to resolve all of the complex factors driving deforestation. The following CIPEC findings counter three myths about forest protection that are pervasive in the conservation community.

Myth 1: Only Legally Designated Protected Areas Will Maintain Forest Cover

The protected area model is based on the assumption that conservation requires government ownership and regulation. In order to test whether government ownership and protected area regulations are necessary for forest protection, Hayes compared forest vegetation density in 76 legally designated, government-owned protected areas to forest vegetation density in 87 forests that are not legally designated as protected areas. The data about these forests are contained in the IFRI database. The findings demonstrate that both legally defined, government-owned protected areas (which we refer to as "parks") and other types of institutions ("non-parks") that include forests managed by private owners, local forest users, or national government agencies are capable of conserving forests.

A Kolmogorov-Smirnov test compares the distributions of forest vegetation density in the two datasets and can be used to determine if the two distributions differ significantly. The Kolmogorov-Smirnov Z score is 0.472, and the p value for that score is 0.979. As the graph illustrates and the test results confirm, *no* statistically significant difference exists in forest vegetation density between parks and non-parks. In other words, formally protected areas do not have a higher frequency of abundant forest vegetation density than areas with alternative institutional arrangements. Legal designation of protection is by no means a requisite for forest maintenance. Furthermore, government ownership of forest lands is not correlated with higher levels of vegetation density. The IFRI study found no correlation between vegetation density and forest tenure type (private, communal, or public).

A closer look at other CIPEC studies illustrates some of the complex factors that influence forest conservation within and outside protected areas. For example, land-cover change in the Maya Biosphere Reserve in the Petén region of northern Guatemala demonstrates both the ability and inability of protected areas to control deforestation. It also demonstrates the important role that biophysical features and institutional recognition and legitimacy play in forest conservation. The Maya Biosphere Reserve covers more than 21,000 km and consists of four national parks, three wildlife reserves, a multiple-use zone, and a buffer zone. The national parks and biotopes are strict conservation regions in accordance with the recommendation of IUCN category II. The multiple-use zone permits limited extractive forest activities, and the buffer zone permits sustainable forest use and agricultural practices.

The four national parks in the Maya Biosphere Reserve show mixed conservation results. An analysis of Landsat images from 1986, 1993, and 2000 shows that two of the national parks, El Mirador–Río Azul and Tikal, have very little deforestation and remain almost intact. In contrast, Sierra del Lacandón National Park and Laguna del Tigre National Park show signs of substantial deforestation between 1986 and 2000.

The biophysical attributes of the region as well as the institutional history of the creation of the four national parks within the Maya Biosphere Reserve are some of the factors that contribute to the stability of forests in El Mirador–Río Azul and Tikal. In El Mirador–Río Azul, biophysical features contribute to forest protection. The national park is located in the remote, northern region of the Maya Biosphere Reserve and is accessible only by a three-day trip by mule or helicopter. The park is officially managed by the Guatemalan Institute of Anthropology and History and the National Protected Areas Council. Nevertheless, it appears that the remote location of the park may be responsible for the forests' protection rather than its institutional designation as a protected area.

Widespread recognition of the ecological and cultural value of Tikal National Park, combined with strong institutional and financial support, promote effective conservation. Created in 1955, Tikal was well established before the region experienced a dramatic advance of the agricultural frontier in the 1980s and before the creation of the Maya Biosphere Reserve in 1990. Tikal is nationally and internationally acclaimed for its cultural and biological uniqueness. It is also financially successful. Revenue from entry fees not only covers the park's entire budget, but also contributes to the Guatemalan Ministry of Culture and Sports. As a result, Tikal is one of the best-staffed protected areas in Guatemala with well-paid park guards and park officials who are held accountable for the continued protection of the park.

Laguna del Tigre National Park and Sierra del Lacandón National Park have been far less successful in preventing deforestation. These parks do not have the biophysical and institutional advantages that El Mirador–Río Azul and Tikal have. Unlike El Mirador–Río Azul, these parks are located in the accessible southern region of the reserve on the edge of the buffer zone. And, whereas Tikal was well established and recognized as a site of national heritage before farmers and ranchers moved to the region, Laguna del Tigre and Lacandón were created as part of the Maya Biosphere Reserve *after* an aggressive colonization policy established by the Guatemalan government encouraged farmers and ranchers to migrate to the region in the 1980s. The parks have since had to try to mitigate the effects of a previous policy that encouraged agricultural expansion into the forested lands.

The present protected area policies are not able to stave off expansion. The numerous light grey and white patches showing the most recent forest cuts demonstrate that these parks have been unable to restrain farmers, ranchers, and loggers from pushing deeper into the forests. In Laguna del Tigre, the underpaid and understaffed park rangers are unable to prevent illegal activities within the park's borders.

A recent study of the area within and surrounding the Royal Chitwan National Park in the terai region of Nepal also challenges the presumption that effective conservation is likely to occur primarily in government-owned, protected areas. Nagendra, Southworth, Tucker, Carlson, Karmacharya, and

Karna have analyzed regeneration patterns across a time series of remotely sensed images of the Chitwan Valley. They found that the buffer zone accounts for a major part of the regeneration occurring in this landscape. While there is considerable donor-assisted funding to the buffer zone to encourage regeneration, the buffer zone lies outside the park and is not a legally defined protected area.

The findings from the cross-national study, the in-depth case study of the Maya Biosphere Reserve, and the recent study of Royal Chitwan National Park illustrate that while parks do provide forest protection in some settings, their performance varies substantially across sites. Parks are not always effective, nor are they necessarily better at forest conservation than other institutional alternatives.

Myth 2: Top-down Enforcement of Protected Area Rules is Necessary to Protect Forest Cover

In *Requiem for Nature*, John Terborgh argues that communities cannot be left to govern themselves. He stresses that local communities are unable to manage their resource systems and that enforcement must not be voluntary. According to Terborgh, "[a]ctive protection of parks requires a top-down approach because enforcement is invariably in the hands of police and other armed forces that respond only to orders of their commanders."

No doubt exists that monitoring and enforcement are critical in forest management. Studies of protected areas have consistently found monitoring and enforcement correlated with effective conservation management. Similarly, common-pool resource management scholars have found that clearly defined boundaries, monitoring, and a system of graduated sanctions are important components of sustainable resource management systems. Common-pool resource scholars, however, have not found that monitoring and enforcement must be administered by a third party. CIPEC studies of forest management underscore the importance of monitoring and enforcement, but challenge the belief that local resource users cannot monitor and enforce forest management rules.

Work by Gibson, Williams, and Ostrom reinforces the findings that rule monitoring and enforcement are critical for forest protection. Their findings also demonstrate that forest users can enforce forest rules. Gibson and colleagues analyzed the rule-monitoring behavior of 178 forest user groups located in 12 countries and coded them in the IFRI database. The user groups included in this study vary substantially in their organization, level of activities, and age. Seventy-five of the user groups included in the study were substantially organized—they elected their own officials, held regular meetings, and undertook joint activities. On the other hand, 29 of the user groups did not undertake any collective activity with regard to the forest they used—they simply shared similar legal standing in relationship to a forest and harvested forest products for household or commercial purposes. The other user groups varied between these two levels of organization.

In our group interviews with members of these groups, we asked them to report on the regularity with which participants in a group monitor or sanction others' rule conformance. We also obtained measures of the group's social capital, their dependence on forest resources, and their assessment of forest abundance. In general, we found a strong correlation between the level of user group monitoring and forest condition (assessed by the users themselves as well as measured by foresters). We found that the level of user group monitoring and enforcement was correlated with assessments of forest condition even when we controlled for the level of social capital in a group, whether a group was formally organized or not, and whether a user group was heavily dependant on a forest. Thus, not only do some user groups monitor each others' activities for conformance with rules, the level of such activities is positively associated with better forest condition.

Detailed case studies of forest monitoring and enforcement suggest that protected areas may be more effective when they complement local rule enforcement mechanisms. In Uganda, CIPEC colleagues Abwoli Banana and William Gombya-Ssembajjwe found that protected area policies that acknowledge forest dwellers' use rights encourage the dwellers to collaborate with forest officials in order to protect their forests. Banana and Gombya-Ssembajjwe compare forest condition in four government forests: Lwamunda, Mbale, Echuya, and Bukaleba. Lwamunda, Mbale, and Echuya are all roughly 1000–1200 hectares. Bukaleba is larger and encompasses 4500 hectares. All of the forests are managed by the Ugandan Forest Department, which is governed by centralized state policies and characterized by a lack of local participation and insufficient human and financial resources.

The critical difference in management policies is that in Echuya the Forest Department staff allows an Abayanda pygmy community to live in and appropriate products from the forest on a daily basis. All other forest users are only allowed to enter the forest once per week. In their study, Banana and Gombya-Ssembajjwe find less degradation and illegal activities in Echuya than in Lwamunda, Mbale, and Bukaleba forests. The authors attribute Echuya's success, in part, to the monitoring activities of the Abayanda pygmy community who report violations to the Forest Department staff. The authors note that the physical layout of the park also helps protect it, as only one road passes through the forest.

Similarly, a CIPEC study of forest management in Rondônia, Brazil, found greater levels of forest protection when protected area policies coincided with forest users' local institutions than when forest protection policies alienated local traditions. In the 1980s, Brazil's Institute of Colonization and Agricultural Reform established two adjacent colonization projects in northeastern Rondônia, each with distinct property rights systems and architectural design.

One project, Vale do Anari, followed the traditional colonization model that laid out an orthogonal road design in which each farmer was assigned 50 hectares of land, of which 50% was to be preserved as forested land and 50% could be used however the settler desired. In contrast, the other colonization project, Machadinho d'Oeste, was laid out to create 16 forest reserves and developed the settlement based on the area's topography. In this settlement, farmers were able to use their land however they liked so long as they respected the reserve forests. Rubber tappers had long lived in these forests and were given rights to help devise a management plan and use the reserve forests. In his research, Batistella found that the rubber tappers soon became unofficial, but active, reserve monitors.

The success of the Machadinho d'Oeste project compared to the Vale do Anari project is illustrated by Landsat images that show the percentage of forest cover loss over time. Images from 1988, 1994, and 1998 show that in 1988, during the initial implementation of the projects, both settlements had similar percentages of forest and pasture. However, by 1998 the overall rate of deforestation in Anari was consistently higher than in Machadinho. The satellite images demonstrate that, while individuals in both settlements deforested lands for pasture and agriculture, the reserve boundaries in Machadinho were maintained and less forest fragmentation occurred in Machadinho than in Anari.

As the individual case studies and cross-national investigation confirm, forest users are able to monitor and enforce forest regulations. In fact, as the Uganda and Brazil examples demonstrate, when local institutions and participation are carefully considered in protected area policies, forest users can be crucial in the invocation and application of forest protection policies.

In some settings, "nature," rather than government officials, protects a park. The remote location of a protected area may be largely responsible for its preservation. El Mirador–Río Azul in the Maya Biosphere Reserve is only accessible via helicopter or a three-day mule trip. This park is apparently in very stable condition even though little planning has been invested in monitoring and enforcement of its boundaries. Similarly, Southworth, Nagendra, Carlson, and Tucker have found that the core areas of Celaque National Park in western Honduras are largely protected due to their inaccessibility and location at elevations above 2300 meters. Many members of local communities surrounding the park are not aware of the park's existence or, if they know that a park is in the region, they do not know the exact location of the boundaries, and little investment has been made in monitoring and enforcement. The recent expansion of coffee production and agriculture, however, has generated considerable pressure on the park's boundaries. Thus, the preservation of the core of this park is not the result of effective management by Honduran government officials.

Myth 3: Local People Are Unable to Make Appropriate Rules

Counter to the presumption that local people are either unable or unwilling to identify forest management requirements and make appropriate rules, IFRI studies of forest management show that resource users are capable of crafting forest rules. Research on the correlation between forest product rules and forest vegetation density in more than 80 IFRI forests in 13 countries finds that the right of user groups to define the forest rules is significantly correlated with forest vegetation density at the 0.05 level.

In the study, rule-making abilities are compared between forests that have above-average vegetation density to forests that have below-average vegetation density, as assessed by the independent forester or ecologist who had completed the forest mensuration. Forest vegetation density is sparser in forests where users do not have the right to define the forest rules and higher in forests where they have rule-making responsibilities. In 24 of the 41 forests ranked as having below-average vegetation density, not a single user group has rule-making responsibilities. In contrast, in 24 of the 43 forests considered to have above-average vegetation density, all user groups participate in forest rule making.

The majority of the protected areas in the IFRI study do not give local people the right to make forest product rules for the forests they use. Seventy percent of the non-parks permit all user groups to participate in the forest rule making, compared to only 22% of the parks. Granting rule-making rights to local forest users increases the likelihood that there will be forest product rules to regulate forest use. Investigation of rule making in IFRI forests finds the presence of forest product rules positively and significantly correlates with the ability of user groups to make rules. For example, in 25 of the 39 forests where all user groups are able to make decisions, rules for all forest products have been established by these groups. In only 1 of the 39 forests where users are able to make rules did users decide not to make any rules for the forest products they use. The presence of forest product rules is also significantly correlated with higher levels of vegetation density. Protected areas, however, do not promote the creation of forest product rules. The officially protected areas in the IFRI database have less than half the number of rules than non-parks.

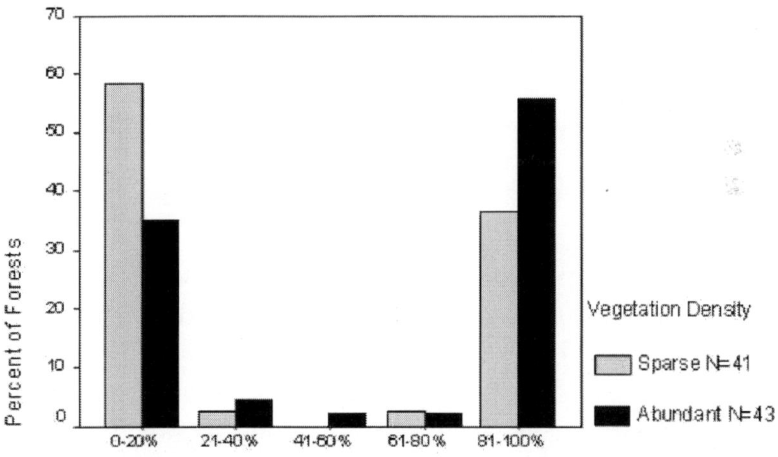

Figure 1. Vegetation Density Associated with User Group Right to Make Rules.

A promising finding from the IFRI study is that parks that give local users rule-making rights have higher levels of vegetation density than parks that do not allow users to make rules. As Figure 2 demonstrates, the parks that do not give any of the local users rule-making abilities have significantly sparser forests than the parks that grant local rule-making responsibilities. A Spearman's rho correlation coefficient of 0.345 confirms a positive correlation between user group rule-making rights and forest vegetation density that is significant at the 0.05 level. These results echo the findings in Uganda and Brazil that inclusion of local forest users and their institutions in protected area planning and administration may complement park policies for pro

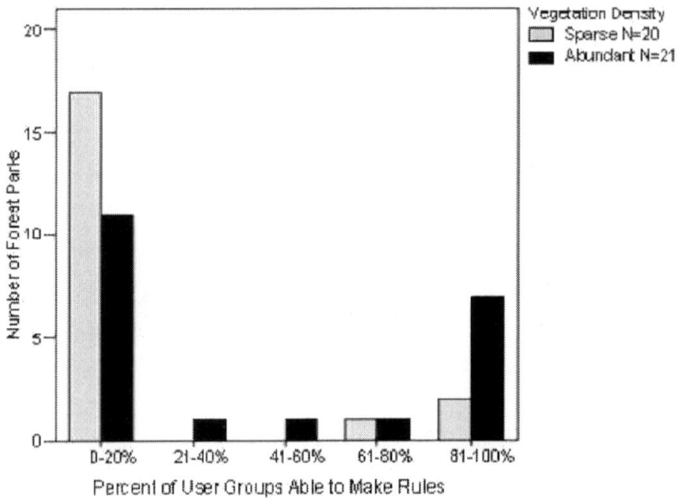

Figure 2. Forest Park Vegetation Density Associated with User Group's Right to Make Rules

Policy Implications

The above findings have several policy implications for forest management, specifically protected area planning and policy design. First, the findings from the IFRI study on protected area forests and several of the case studies clearly refute the belief that *only* legally designated protected areas conserve forests. The findings suggest institutional legitimacy, biophysical features, and the perspective of local residents may greatly influence the ability of a protected area to retain the forest cover within its borders. The findings from case studies as well as the comparative IFRI studies all emphasize the importance of local recognition and respect for protected area policies.

Second, local participation in rule making, monitoring, and enforcement are consistently shown to be significant factors in protecting forest cover.

In Uganda and Brazil, protected area administrators increased the legitimacy of the protected area policies and decreased monitoring costs by granting certain local users forest access and use rights in addition to rule-making and monitoring responsibilities. IFRI results show that local users who have been granted rule-making rights have forests with higher levels of vegetation density than those who do not have those rights.

Finally, a topic of constant debate in the conservation arena is over the level of participation that resource users should have in protected area rule-making and management. Few systematic studies have documented whether local participation makes a difference in forest conservation. The results from the IFRI study on protected areas finds that protected area policies that enable residents to make forest rules are more likely to maintain forest cover. This is an important finding for all involved in protected area planning and implementation.

The above findings clearly contradict the belief that protected areas are the only way to conserve forests. While legal designation of protection may contribute to a basic institutional infrastructure that supports conservation, legal designation alone is never enough to ensure forest protection. As the empirical findings from the IFRI studies demonstrate, protected areas do not have higher levels of forest vegetation density than forests that are not legally designated as protected areas. Similarly, the composite image of forest-cover change in the Maya Biosphere Reserve in Guatemala clearly shows that protected area policies do not ensure forest conservation. Instead, forest conservation depends on a web of factors, including, but not limited to, local recognition of the validity of the protected area policy, biophysical features, financial and human resource support, and mechanisms for conflict resolution.

One of the most significant lessons from the empirical studies discussed above is the importance of understanding and recognizing local-level institutions in forest conservation. Local resource users' rule making and monitoring and enforcement activities are significantly and positively correlated with abundant vegetation density. Protected area policies can provide a broader institutional framework, but it is often the local resource users who determine whether conservation policies will be successful. The IFRI forest study shows that, on average, protected areas that do not allow forest users to make rules are ranked lower in vegetation density. The case studies from Uganda and Brazil show forest conservation in protected areas to be, in part, dependent on the monitoring and enforcement activities of the local forest users.

We still have much to learn about policy tools and legal mechanisms for forest conservation. The above findings take a nuanced approach to forest management and begin to dig into the specific policies and legal rights that may promote forest conservation. Nevertheless, these findings contradict the presumptions that resource users are incapable of crafting and enforcing forest management rules. They also contradict the belief that strictly regulated protected areas are the *only* way to ensure forest conservation. Instead, they suggest that a system of rights and conservation policies that link state and local conservation efforts may lead to greater protection of the world's forests.

References

Chape, S. Spalding, M. Jenkins M. *The world's protected areas: status, values and prospects in the 21st century*. Univ de Castilla La Mancha, 2008.

Kevin Bishop et al., *Protected For Ever? Factors Shaping the Future of Protected Areas Policy*, 12 Land Use Policy 291. 1995.

Lund, H. Gyde. *Definitions of Forest, Deforestation, Afforestation, and Reforestation*. Gainesville, VA: Forest Information Services. 2006.

Michael J. B. Green & James Paine, State of the World's Protected Areas at the End of the Twentieth Century, Paper presented at IUCN World Commission on Protected Areas Symposium on *Protected Areas in the 21st Century: From Islands to Networks,* Albany, Australia. Nov. 24–29, 1997.

7

Rights-based Approaches to Environmental Conservation

The conservation of biodiversity and natural resources afects, and is in turn afected by, the realisation of human rights. Conservation can help realise substantive human rights, such as those to health, culture and food. Likewise, the realisation of human rights can create an enabling environment for achieving conservation objectives. However, certain conservation approaches and measures, such as economic or physical displacement of communities, or harsh enforcement mechanisms, can also undermine or violate human rights. The relationships between conservation and human rights are also shaped by history and their broader governance systems.

The conservation community has responsibilities and opportunities to respect human rights and to contribute to their protection and further fulflment. Conservation organisations can adopt appropriate rights-based approaches (RBAs). These involve integrating rights norms, standards, and principles in policy, planning, implementation, and evaluation to help ensure that conservation practice respects rights in all cases, and supports their further realisation where possible. While a promising way forward, applying RBAs to conservation is also likely to present substantial challenges. RBAs are neither silver bullets nor standalone solutions to balancing diverse local and global needs and interests.

Understanding Rights

Rights can be understood as norms and entitlements that create constraints and obligations in interactions between people or institutions. *Human rights* refer to norms that help to protect all people from severe political, legal, social, or other abuses. Tey are based on the understanding that all people are, by virtue of being human, inherently entitled to minimum standards of freedom and dignity, regardless of nationality, place of residence, gender, origin, colour, religion, language, or any other status. Human rights are often, though not always, recognised and expressed in national or international law. Customary rights are also important, and may or may not be recognised in such legal frameworks.

Despite clarifcation and expansion of recognised rights over the last 60 years, their nature and scope remains contested. Tose recognised in international (and much national) law today include:

- *procedural rights,* such as to participate in decision making, acquire information and access justice; and
- *substantive rights,* such as to life, personal security, health, an adequate standard of living, education, freedom to practice culture and freedom from all forms of discrimination.

Cross cutting *rights principles,* or governing characteristics, include:

- *Universality and inalienability*: All people everywhere in the world possess rights and they cannot be taken away.
- *Indivisibility*: Civil, cultural, economic, political and social rights all have equal status and all must be recognised for human dignity.
- *Interdependence and interrelatedness*: The realisation of one right often depends wholly or in part upon the realisation of others.
- *Equality and non-discrimination*: All individuals are equal and are entitled to their human rights without discrimination of any kind.
- *Participation and inclusion*: Every person and all people are entitled to active free and meaningful participation in, contribution to, and enjoyment of, governance systems in which human rights and fundamental freedoms can be realised.
- *Accountability and rule of law*: States and other duty bearers have obligations to observe human rights, and are answerable for the observance of rights under their jurisdiction.

The range of human rights instruments, and their often high levels of ratifcation, demonstrate wide international acceptance. However, rights norms and principles are not without controversy. Some challenge universality on the grounds that human rights refect western cultural traditions, or more generally that universality is difcult to defend given global cultural diversity. The principles of indivisibility, inter-dependence and interrelatedness gained wide acceptance after the 1993 World Conference on Human Rights. This was connected to a parallel expansion in recognition of the importance of cultural, economic and social rights.

A key feature of human rights is their corresponding responsibilities. All human beings are rights holders. The individuals and groups responsible for upholding and enabling the realisation of rights are duty bearers. States are primary duty bearers, but other non-state actors also have important responsibilities (as discussed in more detail below). Duty bearer responsibilities are typically categorised as respecting, protecting and fulflling rights.

- *Respecting rights* means refraining from interfering with people's pursuit or enjoyment of their rights, for example through uncompensated or forced eviction.
- *Protecting rights* means ensuring that 'third parties' (including private businesses and NGOs) do not interfere with people's pursuit or enjoyment of their rights.
- *Rights-based approaches* Exploring issues and opportunities for conservation
- *Fulflling rights* means creating an enabling environment for people to realise their rights. Rights must be directly provided when people cannot provide them for themselves, such as providing food aid following a severe drought.

Within the UN framework, the obligation to respect is taken to be immediate. Fulflment, however, can be realised progressively, and in line with maximum available resources. Many human rights instruments also allow for a 'margin of discretion', which allows duty bearers to tailor eforts to meet obligations to their context. However, in all cases forward progress must be made, all measures must at all times be non-discriminatory, states must commit the maximum possible resources to fulflment and certain actions (such as removing unnecessary legislative barriers) must be enacted immediately. Priority should be given to securing a 'core minimum', that is, 'ensur[ing] the satisfaction of, at the very least, minimum essential levels of each of the rights.'

To clarify the specifc implications of human rights, for example, for an 'adequate standard of living', a number of detailed standards and indicators are included in human rights instruments, supporting guidelines and other sources. The UN Treaty Body, 'General Comments' provide particularly detailed explanations of standards demanded by certain instruments. The Committee on Economic, Social, and Cultural Rights General Comment 15 on the right to water for example, provides detailed implementation and monitoring criteria, and sets a precedent for the treatment of similar resource rights.

Adapted from CESCR General Comment 12 (para. 7–13) and other sources as cited International law instruments recognising the right to food include: the International Covenant on Economic, Social and Cultural Rights (ICESCR) Article 11, the Universal Declaration of Human Rights (UDHR) Article 25, and the Convention on the Rights of the Child Articles 24 and 27.

The right to food is both a fundamental right to be free from hunger and a key component of the 'right to an adequate standard of living'. 'Adequate' food includes the following:

- *Availability* of foods that are *culturally acceptable, free from adverse substances* and of a *quality and quantity sufcient to meet dietary needs*. Availability refers to the possibilities either of feeding oneself directly from productive land or other natural resources, or of well functioning systems that can move food to where it is needed.
- *Economic accessibility,* which implies that personal or household (financial and other) resources from livelihood strategies, wages and other economic activities should be sufcient for the acquisition of food. This food should support an adequate diet without interfering with the attainment and satisfaction of other basic needs.
- *Physical accessibility*, which implies that adequate food should be accessible to everyone, including physically vulnerable individuals, such as infants and young children, elderly people, the physically disabled, landless people, indigenous peoples, migrants and refugees, victims of natural disasters and other vulnerable groups.
- Availability and accessibility should be ensured through systems that are *ecologically and economically sustainable* and do *not interfere with the enjoyment of other human rights*.

Collectively, these elements also form the pillars of food security.

Recognition of Environment

Recognition of the links between the enjoyment of human rights and environmental protection, broadly speaking, has been growing within the UN, government, civil society and other sectors for several decades, starting with the 1972 Stockholm Declaration. Many international human rights instruments and multilateral environmental agreements now recognise rights to participation in environmental decision making, the importance of the environment for sustainable development and substantive rights to a clean and healthy environment. At the national level, over 100 national constitutions, and nearly all constitutions adopted since 1992, explicitly recognise rights to a clean or healthy environment. Many national and regional courts have also interpreted other human rights, such rights privacy and health, as requiring certain environmental protections. Many civil society organisations now focus on environmental justice issues. Finally, discussion on the human rights implications of climate change and its mitigation is now rapidly increasing.

Specifc attention to links between human rights and *biodiversity / natural resource conservation* has been slower to emerge, but is gaining momentum. According to Alcorn and Royo, human rights are already 'a smoking gun issue' for the conservation community. The literature documents synergies between rights and conservation. It also catalogues instances of rights infringements arising from conservation, including economic and physical displacement. Organised movements of indigenous peoples, local communities and other civil society actors increasingly demand greater accountability from conservation actors regarding past and present results.

Many human rights instruments and multilateral environmental agreements within the UN framework have rights and justice provisions relevant to conservation practice. The CBD, for example, calls for equity in access and beneft sharing from the use of genetic resources with indigenous peoples and local communities, including in the context of protected areas.

Relationships between Rights and Conservation

Drawing on case studies, informal interviews and a review of the literature on the social dimensions of conservation, several closely interrelated issues become clear.

- Conservation can help realise rights through, among many other things, securing sustainable natural resources and ecosystem services to support human health and an adequate standard of living. Likewise, conservation outcomes can be enhanced where people's rights and access are secure, including collective rights to lands and resources.
- Certain conservation approaches and measures, however, can undermine or violate human rights, including through economic or physical displacement from lands or resources important to livelihoods and culture.
- Cases of oppressive conservation enforcement measures, including around protected areas, also raise concerns about personal security rights in conservation practice.
- Failure to conserve ecosystem goods and services can undermine human rights by eliminating people's means of subsistence or destroying essential services, such as those provided by mangrove stands which protect people from the efects of storms and waves; and
- Failing to fulfl human rights can lead to environmental destruction by reducing people's options for the sustainable realisation of their own basic needs and interests.

In the highlands of Peru, six communities of the Quechua peoples have established a Potato Park (Parque de la Papa) to preserve the role of indigenous biocultural heritage (IBCH) for local rights, livelihoods, conservation and the sustainable use of agro-biodiversity. The Potato Park is dedicated to safeguarding and enhancing Andean food systems and native agro-biodiversity. Over 8500 ha of titled communal land are being jointly managed to conserve about 1200 potato varieties (cultivated and wild) as well as the natural ecosystems of the Andes. The villages entered into an agreement with the International Potato Center to repatriate 206 additional varieties, and have a long-term goal to reestablish all of the world's 4000 known potato varieties in the valley.

The Potato Park was initiated by an indigenously run organisation, the Quechua-Aymara Association for Sustainable Livelihoods (ANDES). The initiative has brought together communities that had experienced land conficts, partly through the revival of the village boundary festival in which the boundaries are 'walked'. The traditional knowledge, innovations and practices of the Quechua people are showcased in the Park. Traditional techniques are being augmented by new ones, including greenhouses, education on potato varieties through videos in the local language and production of medicines for local sale. Native species are also being used to regenerate forests and a form of 'agro-ecotourism' is being developed. Geographical indicators and trade marks are used to ensure protection of collective rights and access to markets.

The Potato Park is a community led and rights-based approach to conservation. It is concerned particularly with the Quechua peoples' self-determination, security of tenure and rights to agricultural biodiversity, local products, traditional knowledge, and related ecosystem goods and services. This approach helps to secure local livelihoods using the knowledge, traditions and philosophies of indigenous people.

Economic and physical displacement related to protected areas is one of the more contentious and increasingly visible conficts between conservation and human rights. Such displacements have been reported world wide, though the data related to these reports varies in quality.

In South Africa and Botswana, the contemporary conficts over park-related displacements that took place decades ago are still being played out, and new eforts at redress are being made. After the establishment of the South African Kalahari Gemsbok Park in 1931, the Kalahari San peoples' rights to live and hunt on the land were gradually eroded until they were fnally evicted from the park in the mid 1970s. Similarly, since the start of the colonial era in Botswana, the establishment and management of protected areas has often involved forced and uncompensated displacement of people, especially the minority peoples often referred to as Khoi San groups. These displacements, which deprived people of the basis for their livelihood, can be seen as a violation of rights to customary lands, an adequate standard of living, health, culture, freedom from discrimination, self-determination and a host of procedural rights.

In recent years, some of the displaced communities have made eforts, and had partial success, in getting redress and restitution of their rights. In Botswana, a San group successfully challenged a recent eviction associated with the Central Kgalagadi Game Reserve and partially won the case. The San were granted the right to return to their settlement in the game reserve. What remains a human rights challenge is that the government is not obliged to provide the community with social services, as they do in other settlements. Similar and closely related changes have been attempted in

South Africa under the new democratic government. In 1999, after four years of negotiation, the 'reconstituted' Khomani San community land claim was partially successful, with the restitution of ownership of 25 000 ha of park land and shared management rights through a Joint Management Board. The San can now use this land in accordance with conditions contained in the settlement agreement and the management plan. No permanent residence, agriculture or mining are allowed, but tourism related and traditional/cultural activities are, including hunting, provided such use and activities are sustainable and in keeping with biodiversity conservation objectives. They have only cultural and symbolic rights to the remaining 350 000 ha they claimed. They were also granted six farms outside the park, but do not have the same cultural connection to this land. Between 1999 and 2002 there were additional negotiations to settle a conficting land claim submitted by the neighbouring Mier community, which was eventually awarded 25 000 ha. These cases refect just some of the vast complexity and competing interests and claims that can be involved in rights restitution.

Concepts and Components for RBAs to Conservation

Rights-based approaches, increasingly adopted in development, business, forestry and other sectors, may hold great promise for conservation. But what are the practical implications of such approaches? What are their benefts and challenges, and for whom?

There is no consensus on the defnition or form of RBAs. RBAs can be understood as integrating rights norms, standards, and principles into policy, planning, implementation, and outcomes assessment to help ensure that conservation practice respects rights in all cases, and supports their further realisation where possible. It often includes eforts to make human rights and conservation mutually and positively reinforcing pursuits.

The conservation community can learn from experience in other sectors. RBAs to development generally focus on *enhancing* human development through protecting and fulflling rights, particularly for the most vulnerable. In the business sector, RBA models often involve measures to avoid infringing on people's enjoyment of their rights by creating internal safeguards, but also by developing supply chain control mechanisms to help ensure partners also respect basic rights. Both safeguarding against infringements and pursuing further rights realisation are relevant to RBAs to biodiversity and natural resources conservation. The conservation community may also need to develop new dimensions of RBAs to refect concerns particular to their arena, for example, inter-generational rights, environmental sustainability and governance of natural resources.

RBAs can be carried out at multiple scales and contexts, and through various legal instruments, policies, programming approaches, methods and tools. Key elements can include:

- identifying all relevant rights claims and obligations, including customary and collective rights;
- using rights norms, standards and principles to guide policy, programming and implementation;
- analysing and monitoring processes and outcomes against rights-based criteria;
- engaging with the rights implications of conservation practice as a matter of *obligation;*
- supporting eforts to address the underlying causes of rights violations, including by changing inequitable power relations;
- building the capacity of both rights holders and duty bearers to claim their rights and meet their respective responsibilities;
- taking all available measures to respect rights in all cases and supporting their protection and further fulfilment wherever possible, particularly for the most vulnerable; and
- supporting eforts to provide access to justice and redress for violations, where needed.

Examples of RBAs to development, business practice and conservation are typically framed as approaches carried out by duty bearers or third parties. Tey often include, among other things, measures to assist rights holders and duty bearers in exercising their rights and meeting their obligations. At the same time, rights holders often have extensive knowledge and capacity for defning and asserting their rights and their advocacy actions have often expanded rights recognition. Case 1 illustrates how communities used their conservation knowledge and strategies to enhance their cultural, food, land and other rights.

Accountability in RBAs to Conservation

RBAs to conservation raise challenging questions about who the rights holders and duty bearers are, how they relate to one another, and the nature and scope of their rights and responsibilities. States ratify international human rights instruments and are responsible for implementing national laws that uphold those obligations. However, third states (states acting within or afecting another state) and non-state actors, including private businesses, international organisations and NGOs, can have substantial positive and negative infuences on rights.

There is growing recognition that states should take responsibility for the human rights efects of their policies in other countries, particularly regarding the obligations to respect and protect. Non-state actors should also be held accountable for their efects on rights, even where not directly regulated by a state. '... [I]n an age when other public and private actors are more powerful than states, human rights must be extended to limit their potential abuses of power against people. ... With power must come responsibility.' While referring primarily to transnational corporations and international organisations, the sentiment of Ziegler's statement is relevant for many non-governmental conservation actors. Contemporary discussion about NGO accountability and calls for NGOs to respect human rights will likely only increase. James Igoe, for instance, argues that the increasingly powerful role of non-state conservation actors(private and NGO) in eastern Africa must carry increased human rights responsibility, though continued and ultimate responsibility lies with the state. Non-state accountability should neither weaken the state's role, nor suggest that states are absolved of their responsibilities, because the state provides the institutional, policy and legal frameworks within which civil society exists and exercises its rights.

What do these state and non-state obligations imply in practical terms for RBAs to conservation? RBAs would require taking explicit measures to respect rights—not infringing on people's enjoyment of their basic rights and avoiding harm to vulnerable people—in all cases. In this sense, restricting the access to or use of natural resources in cases where people rely on those resources for their livelihoods, without providing a viable, sustainable and better alternative, would be considered a failure to respect their rights. Beyond this, conservation organisations can, within their (often substantial) spheres of infuence, help protect rights. This could involve encouraging and building the capacity of partners and other duty bearers to respect rights, for instance by training park guards on the use of rights consistent enforcement measures. An organisation might also choose to put specifc obligations in contractual arrangements with partners regarding their rights-related actions. In working with rights holders, conservation organisations can also provide information to local people to help build their capacity to protect and demand their own rights. Finally, under 'supporting fulfilment', conservation organisations can take proactive measures to make conservation 'work for rights'. This could mean moving far outside the scope of the organisation's normal sphere, such as using revenues from conservation to build schools in communities near parks. However, supporting fulfilment of rights can also involve improving management of a degraded watershed to provide important, additional food and water services to local people. This kind of improvement will, in turn, help people fulfl their rights to health, water and an adequate standard of living (assuming they retain access to those services). The same actors may sit at diferent places along this continuum with respect to diferent rights. For instance, in some cases *respecting* the right to food access and availability could best be achieved by supporting the *fulflment* of procedural rights and security of tenure/resource access. In all cases, the distinctions and linkages between 'respecting, protecting, and fulflling' rights in RBAs to conservation—and their concrete implications for practitioners—require further exploration and learning.

Rights holders' responsibilities are another important piece of the RBA puzzle. Everyone has responsibilities not to infringe on the rights of others. This raises challenging questions in the conservation context. How can upstream and downstream communities and individuals both respect one another's rights, and fulfl their own rights, when sharing insufcient watershed resources? Further, how can these communities, already struggling with scarcity, respect the rights of future generations to those same (or similar) resources? Regarding global justice and intra-generational rights, how can dispersed duty bearers be made accountable for the efects on rights of their contributions to environmental destruction, as in climate change? What responsibilities do people in wealthy countries have to use and redistribute resources in ways that enable *all* people to meet their basic needs in sustainable ways?

RBAs to conservation also raise difcult questions about rights holders' responsibilities with respect to biodiversity and non-human rights. How can RBAs to conservation both respect the inalienability of rights and support restricting use of natural resources for conservation, particularly where poor or vulnerable peoples are using those resources in support of their livelihoods or culture? The 'key question is how to assign responsibilities fairly and efectively … while maintaining an overall rights-based approach'. In all cases, this 'fair' distribution of responsibilities should respect basic procedural and substantive rights. In principle this seems straightforward. In practice, it may be a difcult position to navigate, particularly in situations where resources are scarce or threatened.

RBAs cannot resolve all of these complex dilemmas. Tey can, however, provide a useful framework by helping to systematically identify and analyse the key issues, design a process for negotiating competing claims and resolve grievances. RBAs can also identify the minimum standards that should not be 'traded-of' in negotiations.

Conservation, Rights, and Governance

Natural resource governance, and the broader governance systems in which conservation and rights are interacting, can both support and be supported by RBAs. A focus on improved governance should be part of RBAs to conservation, and vice-versa.

Natural resource governance can be understood as 'the interactions among structures, processes and traditions that determine how power and responsibilities are exercised, how decisions are taken, and how citizens or other stakeholders have their say in the management of natural resources—including biodiversity conservation.' While there is no global consensus, commonly recognised elements of 'good' (that is, efective and equitable) governance include: transparency; access to information; access to justice (including a means of resolving confict and disputes); participation, legitimacy, and voice (genuine involvement in decision making); fairness; coherence; performance; subsidiarity; respect for human rights; accountability; and rule of law (fair, transparent and consistent enforcement of legal provisions). The overlap between elements of RBAs and 'good' governance—respect for human rights, as well as fairness, meaningful participation, accountability, rule of law, and access to justice—clearly illustrate the close links between them. Additionally, according to the UN OHCHR's RBA guidelines, 'a human rights-based approach … premise [is] that a country cannot achieve sustained progress without recognising human rights principles (especially universality) as core principles of governance. … *The concepts of good governance and human rights are mutually reinforcing…*' .

Taken another way, the 'interactions among structures, processes and traditions that determine how power and responsibilities are exercised'—governance—are also a major factor in determining how conservation and rights are related in any particular context. It is the broader governance arrangements that shape the way that conservation will afect rights and that also help determine what options are available for addressing those afects. In this way, too, governance and rights in conservation are inextricably linked. *To be efective, then, RBAs must account for and focus on improving the governance systems through which the approach is being carried out.*

Shared governance and community governance may be given special consideration. RBAs work to enhance equity by increasing the capacity of both rights holders and duty bearers to realise human rights. Shared governance is one way to bring rights holders and duty bearers together in transparent processes in which they can understand the claims and duties at stake, and negotiate fair outcomes. Community governance supports collective rights, including the right to self-determination, and empowers communities to mobilise their capacities to protect and fulfl their rights and conservation vision. However, shared and community governance should also refect rights-based approaches, including attention to rights issues within and across communities.

Introducing RBAs as a 'new' approach to conservation begs the question of how they difers from other methods. State and NGO led conservation approaches have evolved over the last several decades from practices which were often exclusionary to ones that increasingly (though by no means always) embrace local people's participation, interests, and knowledge. RBAs can be seen, in part, as a continuation of the progression towards more inclusive and socially just conservation. But what exactly does it add?

By drawing on the human rights framework—a relatively well developed and widely recognised set of negotiated standards—RBAs ofer a strong standard with which to understand and assess the social implications of conservation practice. Further, the substantive and procedural aspects of inclusive RBAs can be far more comprehensive than a more general 'participatory' or 'pro-poor' approach. RBAs can also enhance accountability by linking rights (and rights holders) with specifc, corresponding obligations (and duty bearers), including consideration for particularly vulnerable individuals. By linking rights and obligations at the individual and sub-group levels, in addition to the community/state/NGO levels, RBAs can provide a powerful addition to approaches that fail to adequately address inequities and vulnerability within and across communities.

RBAs may also provide a stronger foundation for incorporating human wellbeing concerns by recognising that doing so is a matter of obligation. The rationale for incorporating people's knowledge and interests in conservation practice—through, for example, participatory, community-based, and decentralised natural resource management regimes—has often been that doing so will enhance conservation outcomes. It does so by drawing on local knowledge, enhancing local ownership, and reducing confict and non-compliance. In other words, an instrumental approach. Such approaches have sometimes been criticised on the grounds that they engage with people only at a superfcial level, and that conservation costs and benefts are not evenly distributed within and across communities, as powerful diferentials can lead to elite capture.

Additionally, where collaborative approaches have proven too costly or difcult, conservation organisations have sometimes moved back towards more overtly protectionist and exclusionary models. Finally, approaches that exclude local people *can* sometimes be efective in achieving conservation outcomes. For all of these reasons, instrumental approaches *alone* may be insufcient to guarantee that people's wellbeing is secured. By addressing human wellbeing as a matter of *obligation*, and addressing the rights not only of communities, but also of individuals and vulnerable groups within communities, RBAs can, in principle, better ensure that basic human rights are respected.

At the same time, it is difcult, and perhaps not very useful, to make too strict a distinction between rights-based and instrumentally-driven approaches. RBAs are not stand alone solutions; they are likely to be one component in a mix of approaches with diferent rationales. Further, as recognised by UN OHCHR, in the context of development, RBAs can make programming more efective. RBAs *can* enhance conservation outcomes, and those conservation benefts can rightly form part of an organisation's motivation for adopting such approaches.

Rights Relevant for RBAs to Conservation

RBA encompass a vast array of potentially relevant rights, recognised in:

- treaties and declarations of the UN;

- regional human rights instruments ;
- national constitutions, law and regulation, which are also often the basis for implementing international law obligations;
- customary law, norms and practices, which may or may not be recognised as legal rights by states;
- contracts, codes of conduct and organisational standards of governmental, non-governmental and community institutions;
- Multilateral environmental agreements (for example, CBD), that, while not rights instruments *per se,* include social standards; and
- other contextually relevant instruments and standards.

Within these sources, relevant *procedural rights* include: information, participation in decision making and access to justice. Relevant *substantive rights* may include (but are not limited to):

- life;
- privacy;
- health;
- culture and religion;
- freedom from hunger;
- freedom from all forms of discrimination;
- development;
- right to (a healthy, safe) environment;
- voluntary, safe and healthy working conditions;
- adequate standard of living (including food and housing);
- access and beneft sharing in the use of natural resources;
- peoples' rights to self-determination, to freely dispose of natural wealth and resources, and not to be deprived of their means of subsistence;
- indigenous peoples' rights to maintaining traditional ways of life, free and informed consent prior to activities on their lands, self-representation through their own institutions, freedom to exercise customary law, and ownership, control, and development of communal lands, territories, and resources traditionally owned or otherwise occupied;
- rights of redress for infringements, including restitution, compensation and satisfaction; and

- other customary or contextually relevant rights.

The human rights framework typically focuses on individual rights holders. It is only recently that established international human rights instruments have addressed, or been reinterpreted to address, collective rights. Collective rights of indigenous peoples and local and mobile communities may be particularly important for drawing out potential synergies between rights realisation and conservation. Strengthening collective land tenure rights, for instance, can help provide incentives and support (customary and new) community institutions for efective local natural resource management.

Finally, there are several emerging rights issues to which the conservation community may be able to contribute understanding and capacity for action, including the following:

- *Inter-generational rights* imply that present generations have a duty to protect and sustainably manage natural resources and the common heritage of humankind. There are difcult questions regarding practical application: Who can represent future generations? How can present generations determine the needs of future generations and ensure respect for their rights? Conservation that helps secure resource sustainability across generations may contribute to the understanding and realisation of inter-generational rights.
- *Climate change* already has, and will likely continue to have, profound efects on the enjoyment of human rights, particularly for the most vulnerable. Climate hazards may result in mass displacement, food insecurity, the spread of endemic disease, water scarcity and other issues. Adaptation is a pressing need, linked to many substantive human rights. Climate change efects also pose major *intra-generational rights* and global justice concerns; the most vulnerable countries and people are also, typically, those who have contributed least to greenhouse gas emissions. There are also concerns about the potential efects on human rights of major climate change mitigation strategies, including Reduced Emissions from Deforestation and Forest Degradation (REDD) and biofuel production.
- '*Environmental rights*', (the right to a healthy environment), *rights-based approaches to environmental protection* (using human rights law to thwart environmental destruction), and *non-human rights* are all of interest to the conservation community.

Considering the range of relevant rights, and the often very limited resources available, must RBAs to conservation address *all* human rights in *all* initiatives? For any given initiative, it may be only a sub-set of rights that are at risk of being undermined, or that can be further fulflled, through conservation. UN OHCHR suggests that RBAs to development should focus on the most pressing rights. However, this requires knowing which rights those are. Tus, *RBAs require up front and ongoing analysis of the specifc rights issues relevant to the context.* In all cases, procedural rights and cross-cutting principles, such as universality, inclusion, accountability and non-discrimination, are important.

International law recognises the rights to popular participation in decision making, information and access to justice among other procedural rights. *Procedural rights are important ends in themselves. They can also be entry points for securing substantive rights.* Information can help ensure that rights holders, duty bearers and other interested parties understand their respective entitlements and obligations. Inclusive processes (participation) that include options for redress or access to justice can help rights holders and duty bearers engage with one another, and hold one another accountable, in negotiations for satisfactory solutions. 'Accountabilities for achieving [RBA] results or standards are determined through participatory processes, ... and refect the consensus between those whose rights are violated and those with a duty to act.' UN OHCHR.

Using procedural rights to identify and secure other rights can also be helpful because *substantive rights may, on their own, be difcult to operationalise or support.* For instance, the right to culture is highly relevant for people whose culture and livelihoods are closely linked to lands and resources being conserved. However, the practical implications of the right to culture are poorly understood (Centre for Economic and Social Rights website). Similarly, based on CARE International's RBA in Uganda, Franks emonstrates that, while protected areas have sometimes had negative efects on substantive rights (for example, food), such rights 'ofer little in the way of practical means of addressing the problem', but that 'working to ensure procedural rights can be an efective entry point for positive social and environmental impacts'.

The principle of universality suggests that RBAs to conservation consider the rights of all stakeholders. At the same time, UN OHCHR states that, '[a] human rights-based approach focuses on the realization of the rights of the excluded and marginalized populations, and those whose rights are at risk of being violated .... Universality means that all people have human rights, even if resource constraints imply prioritization. It does *not* mean that all problems of all people must be tackled at once'. Consistent with this, RBAs to conservation may also focus *primarily* on supporting protection or fulfilment of the rights of marginalised and vulnerable populations, or those facing the greatest (positive and negative) conservation-related efects on human rights. Attention to the most vulnerable within RBAs is intended in part to avoid 'elite capture'. However, this notion raises challenging questions regarding how, and by whom, the 'most vulnerable' are to be identifed, including within and across local and mobile communities. Understanding who the most vulnerable are requires a nuanced, disaggregated picture of the context. This is no simple matter. In all cases, a focus *only* on the most vulnerable, to the exclusion of others' rights and interests, would violate the principle of universality. RBAs should also consider and at least respect *all* people's basic human rights.

Finally, as with all people, conservation actors are also rights holders. Many conservation practitioners have faced human rights abuses as a result of their work. IUCN members at the 2000 World Conservation Congress (WCC) (Amman, Jordan) adopted a resolution—Support for environmental defenders—that calls on IUCN to speak out 'publically and forcefully' in support of environmental defenders facing rights abuse. A further resolution passed at the 2008 WCC calls on IUCN members and other stakeholders to ensure adequate protection of rangers who are defending protected environments, including through appropriate legislative measures.

Respecting and strengthening land tenure and resource access rights can be an important mechanism for harnessing the positive synergies between conservation and human wellbeing. Formal, informal and customary access, use, management and exclusion rights are often intricately connected to incentives for sustainable use.

There are also potential challenges of focusing RBAs on such rights. Tenure rights are highly complex, and fraught with difculties of competing claims, lack of data, lack of political will and other challenges. These, however, are challenges that may be common to many rights.

Another potential problem is that land rights may not necessarily be considered a *human right* in all cases. Instruments concerning indigenous and traditional peoples recognise rights to traditionally owned and occupied territories. Tenure rights can also be recognised in national or customary law, such as many community land and forest laws. However, for many vulnerable people—including the millions of formally landless rural people—a clear 'human right' to land is weak or nonexistent. Here the normative content of other substantive rights, including the right to an adequate standard of living, can support basic land and resource access claims for rural, landless peoples. Similarly, several of the UN voluntary guidelines on the right to food make direct links between food rights and natural resource sustainability and accessibility.

Conservation Approaches

If well designed and governed, certain conservation strategies may be more efective than others at simultaneously meeting conservation and human rights objectives; examples include:

- *landscape and ecosystem approaches* that embrace conservation as a set of processes linked to the broader (cultural, ecological, historical, cultural and political) landscapes;
- *collaborative and community natural resource governance,* including in protected areas, which can institutionalise rights holder participation and mutual accountability, support collective rights, and facilitate communities in acting on their own potential and conservation visions; and
- *natural resource management objectives that include sustainable use,* alone or in combination with other approaches.

However, RBAs to conservation, as conceived of here, do *not preclude* strictly protected areas or any other particular conservation strategy. Rather, they guide and limit how such strategies are developed and implemented. For example, under RBAs, *physical or economic displacement* should be avoided in all cases, for example by seeking all possible alternatives and zoning for sustainable use. More discussion and learning is needed to understand the implications of RBAs where physical or economic displacement is the 'only option' (and to understand under what circumstances displacement can, in fact, be concluded to be the only option). In all cases, though, this would presumably require that duty bearers are held fully accountable for supporting or engaging in a fair, transparent and inclusive process in line with human rights law and all other relevant national law. This would include upholding the free, prior and informed consent of indigenous peoples, and ensuring that afected people are provided with better, culturally appropriate and sustainable alternatives in the short and long run.

Adopting RBAs to Conservation

RBA may present both benefts and challenges for conservation organisations, which will vary depending on the perspectives, interests and capacities of the participants. *Likely* benefts of RBAs, identifed from experience to date in conservation and other felds, include the following:

- improving governance of natural resources and biodiversity, and in doing so potentially enhancing conservation outcomes;
- efectively responding to increasing demands from donors, NGOs, communities, social movements, etc., for greater accountability by conservation organisations for the efects of their activities on rights;
- more fully ensuring the rights of the most marginalised people, in addition to other rights holders and stakeholders, are addressed;
- better addressing possible inequalities within and across communities and levels, i.e. avoiding 'elite capture';
- providing clearer criteria with which to design and measure conservation programming and outcomes by drawing on the widely recognised human rights framework;
- bringing greater analytical clarity regarding the underlying and broader causes of the multidimensional links between conservation and human rights;
- helping demonstrate conservation's positive contributions to human rights, including in emerging areas, such as collective and inter-generational rights, and, conversely, increasing understanding of the rights-related risks of not protecting critical natural resources and biodiversity;
- providing a framework for identifying and balancing competing claims—including global public interests and local people's needs—with attention to the minimum standards that cannot be 'traded-of'; and
- providing a platform for, and guidance on, redress for infringements and violations, where needed.

There are also many *challenges and potential costs* to RBAs to conservation. The human rights framework is by no means simple, and many rights can be difcult to exercise in practice. Further, as discussed above, there may be competing rights claims between vulnerable people relying on the same set of scarce resources. It is difcult within the human rights framework to deal with such conficting claims, or to prioritise among pressing individual rights and broader public interests. There may also be conficts between rights within a given group. Wilkes and Shen, for example, describe how a conservation and food security enhancement programme appears to be undermining the cultural rights of some communities in China by introducing culturally inappropriate and likely, ultimately unsustainable food production methods

RBAs will also require moving outside common boundaries of the conservation arena. Tey demand engagement with the broader political and governance systems that shape the links between conservation and rights. The extent of conservation organisations' infuence over these broader systems will be limited. For example, if there are discriminatory land tenure laws in place, and these laws allow unjust land acquisition (for conservation or other purposes), it may be difcult for conservation organisations alone to counteract the unjust results of the law. At the same time, conservation organisations often do have substantial power. This includes the power to choose not to beneft from unjust laws. An organisation may choose not to work in an area or context where rights violations are occurring, or are likely to occur, in ways that are linked with conservation activities. However, conservation organisations may also have the political infuence to advocate for policy changes, and/or convene processes in which rights holders and duty bearers can discuss and negotiate just arrangements.

Using human rights language can raise new challenges by making explicit the conficts and political tensions underlying conservation. At the same time, human rights can provide a framework for addressing these issues). In all cases, the fact that RBAs pose political challenges is not a justifcation in itself for avoiding engagement.

RBAs require substantial resources—time, expertise (in, for example, human rights law), information (such as disaggregated baseline data and indicators to assess change) and funding—which may be difcult to obtain. Embracing the fact that rights imply specifc *duties* may also require a more fundamental shift in the perspectives of conservation organisations and their staf. Open discussion and better understanding of the ways in which realising human rights supports longer-term nature conservation objectives may assist in this process.

Finally, the timeframe (across generations) and scale (across borders) of conservation's positive contributions may not be easily understood in the traditional human rights framework; a framework which focuses, primarily, on present generation individuals or groups within certain places. The current debates on climate change and human rights are perhaps opening the door for further application of human rights principles to the intra-generational (global) and inter-generational scales.

In the end, the benefts and challenges of RBAs to conservation will be understood only after further learning. The benefts and challenges will also likely vary with the circumstances in which RBAs are adopted and the perceptions and resources of the parties involved.

Failure to conserve resources can contribute to undermining human rights (Large-scale shrimp farming threatens subsistence of coastal communities)

Shrimp farming in Latin American and Asian countries has greatly increased in recent decades, due, in large part, to promotion by governments and international donors who view it as a mechanism for rapid development. However, while shrimp farming is lucrative for the relatively few farm owners, the efect on vulnerable coastal people has been devastating in terms of employment and subsistence livelihoods. Large-scale shrimp farming has led to destruction of wetlands, agricultural land and biodiversity on which the poorest people are directly dependent. Furthermore, the large amounts of shrimp being produced are not priced at local market levels, or otherwise made available to local people. For example, in Esmeralda (Ecuador), approximately 90 000 people (15 000 families) living in fshing villages are threatened by the indiscriminate cutting of the mangrove forest along the mouth of the Muisne River. For several decades, the shrimp industry has been destroying mangroves to set up aquaculture farms and, in so doing, is directly undermining the rights of the 25% of coastal residents who depend directly or indirectly on the mangroves. In sum, unless adequate alternatives or compensation are provided, allowing and promoting large-scale, export driven, shrimp farming can be viewed as a violation of, among others, the rights to health and an adequate standard of living for those people most dependent upon the natural resources destroyed by this practice.

Failing to realise human rights can lead to loss of natural resources and biodiversity (Encroachment and biodiversity loss arising from failure to fulfl the right to food)

In an extensive review of rights campaigns, Vosti and FIAN identify numerous cases in which the failure to fulfl the right to food—for example, failing to address highly skewed land ownership and an acute lack of access to land by the poor— has resulted in encroachment into and destruction of areas of high biodiversity. According to Vosti, 'small farmers … account for about two-thirds of rainforest destruction, by converting land to agriculture … [and] only improvements in agriculture's performance as part of an opening up of alternatives for meeting basic welfare requirements can save the rainforest'.

Among several cases, FIAN documents instances in Brazil and Nepal where severe land access restrictions resulted in migration and encroachment into high biodiversity forests, though in both cases the true percentage of biodiversity destruction attributable to the landless and poor people cannot be determined and is often overestimated. In sum, 'in many (if not most) cases the eco-destruction by victims of violations of the right to food can be seen as resulting more or less directly from the violation itself. Had this violation not occurred, the deprived people would not have had to invade the forest, farm on steep slopes or overgraze fragile pastures'. The implications of these accounts are that efective and long-term protection of biodiversity requires the sustainable fulfment of human rights, including the right to food.

As with other cases, the links between rights and conservation issues are embedded in history and highly complex social, economic and political circumstances.

Developing and Implementing RBA to Conservation

Ultimately, any organisation will have to develop RBAs appropriate to its mandate, resources and objectives. It is neither possible nor desirable to provide a blueprint. However, from the preceding analysis and experiences elsewhere, some *preliminary* recommendations can be identifed. These are general points, which would require further development and adaptation by any conservation organisation. Tey are presented across three categories:

1. policy instruments and organisational commitments on rights;

2. guidelines and strategies for integrating rights into programming and practice; and
3. methods and tools for rights-based planning, monitoring and evaluation.

Policies/Commitments Towards Rights

An important frst step may be for a state, agency, or organisation to develop appropriate laws, policies, or organisational commitments on rights and justice in conservation. For example, a state may adopt stronger rights provisions within national tenure or conservation laws. An NGO may adopt a voluntary code of conduct, or other publicly communicated standard. Such instruments could include the following components.

- Recognising that land and natural resource use restrictions can infringe on the basic rights of vulnerable local people and others, commit to taking concrete, transparent steps to help ensure respect for basic human rights in conservation programming and implementation. Where infringements do occur (or have occurred), provide or facilitate access to redress.
- Recognising that conservation organisations can assist and infuence their partners and other stakeholders, commit to helping ensure other duty bearers meet their obligations (i.e., support rights protection).
- Recognising that conservation outcomes can contribute to local livelihood security and human wellbeing, commit to supporting further rights fulfment wherever possible, especially for the most vulnerable people.
- Recognising that conservation and human rights realisation can be mutually supportive, seek opportunities for positive synergies.
- Recognising the indivisibility and interrelatedness of rights, commit to respecting all relevant procedural and substantive rights, including, but not limited to, the rights of indigenous peoples and local and mobile communities, and including customary norms and rights where appropriate.
- Recognising that experience within the conservation community can contribute to the advancement of human rights, including inter- and intra-generational and environmental rights, seek ways to contribute to new learning and action.

- Recognising that rights-conservation linkages are shaped by broader systems of governance and power, commit to understanding and engaging with other actors and systems as needed to fully develop a rights-based approach.

Strategies to Integrate Rights Considerations into Practice

An organisation may have to decide *to what,* exactly, the rights approach is being taken. Is supporting further rights fulflment part of the organisation's mission, or is respect for rights a safeguard measure? How will rights factor into project and programme planning, fund raising, partnership arrangements, advocacy, monitoring and evaluation, etc? Will RBA implementation be top down (headquarters' policy) or bottom up (project and programme design at the local level)? Management questions like these will have to be resolved by organisations adopting RBAs. Further, embracing RBAs may also require a more fundamental shift in how people perceive the relevance and importance of rights to their work. In other words, if RBAs are only a management framework, and not a refection of a deeper ethnical commitment, they risk remaining superfcial. With this and the preceding analysis in mind, the following are some preliminary guidelines and strategies for RBAs to conservation.

Support integration of rights considerations by:

- undertaking a comprehensive organisational review to determine where and how to integrate rights considerations;
- establishing learning and exchange platforms, internally and between multiple stakeholders. This may require providing additional support to ensure rights holders can participate. It can also include linking rights holders with policy makers and other duty bearers across levels;
- establishing dedicated internal or external bodies to provide assistance and oversight in designing and implementing RBAs;
- developing guidelines and recommendations on RBAs for the organisation, its partners and others it can infuence or assist, including policy makers;
- developing, testing and implementing tools to support RBAs in a variety of contexts; and
- providing training and other capacity building exercises for organisation staf, partners and other rights holders and duty bearers, including in collaboration with human rights organisations.

For any programmes or projects that may have positive or negative rights implications, *develop a clear understanding of the context and issues* by identifying:

- all claims holders, including vulnerable groups and individuals;
- all duty bearers—government and non-government—and the nature and scope of their responsibilities. Include the responsibilities of claims holders where relevant;
- contextually relevant rights and institutions, for example, international, national and local legal and customary rights
- pressing rights concerns or vulnerabilities of claims holders, including the most vulnerable;
- direct and underlying causes for rights realisation and rights concerns, including factors in the political, social, cultural and historical contexts;
- the capacities (resources) and needs (gaps) of rights holders and duty bearers for realising rights and responsibilities; and
- possible rights risks or benefits of ongoing or planned conservation activities, taking account of the broader political and governance systems

Use *rights principles and substantive and procedural rights norms and standards* to guide planning, programming and implementation. Some examples include:

- Promote meaningful participation and access to information for all interested parties.
- Support equality, non-discrimination, and inclusiveness of all individuals and people by: maintaining a disaggregated picture of the context. Provide additional resources for otherwise marginalised groups to participate, developing culturally appropriate processes, using language accessible to all parties, and other steps.
- Establish accountability and access to justice by clearly setting out responsibilities and commitments of all parties. Build the capacity of rights holders and duty bearers to identify and act on their respective entitlements and obligations. Encourage third party evaluation and arbitration and include processes for concerns to be raised and addressed, including redress where needed.

- Help ensure that substantive rights concerns are identifed, addressed in processes and measured in outcomes according to all relevant standards (for example food availability and access).

Use human rights *standards* and *indicators* to *monitor and evaluate* outcomes, using disaggregated data to the greatest possible extent. As *cross cutting principles:*

- Engage with the *broader (policy, political, socio-economic) context and governance systems* that shape the way that conservation may afect rights and that, in turn, defne the opportunities for addressing these.
- Focus on *improved natural resource governance* for biodiversity and human wellbeing.
- Recognise, respect and support local people's *capacities, institutions, knowledge and conservation visions.*
- Approach RBAs as processes of *'learning-by-doing'*, rather than as events or tools.

Examples of Methods and Tools to Support Rights-based Planning, Monitoring, and Evaluation

Planning, monitoring and evaluation methods and tools for RBAs have been developed within the development, human health, business and, more recently, conservation contexts. Examples are briefy described here. Strengths and weaknesses of these methods and tools vary, and all would require further adaptation for use by any particular conservation organisation. These tools typically assess particular project or programme risks, but can also build towards more generalised rights-based programming over time.

Rights Checklists and Compliance Assessments

For new or ongoing projects or programmes, an organisation may develop a checklist of questions or issues to be addressed. Relevant examples have been developed by Filmer-Wilson and Anderson and Svadlenak-Gomez. A more comprehensive tool—a *Human Rights Compliance Assessment*—has been developed by the Danish Institute for Human Rights to help businesses to identify areas of (potential) human rights concern.

Human Rights Impact Assessments

Human Rights Impact Assessments (HRIA) have been developed and tested in development, health, business and other felds and examples in. Like other impact assessments, HRIA are generally multistep processes supported by adaptable tools. Common steps may include:

- developing a nuanced understanding of the human rights and conservation situations in the (political, legal, socio-economic, cultural, historical) context;
- identifying the risks and/or developing a shared vision for rights fulflment;
- formulating options and negotiating (risk mitigation/rights enhancement) activities;
- implementing policies/activities in project/programme cycles, using a 'learning-by-doing' approach; and
- monitoring, evaluating, reporting and changing policies/activities as needed.

Each step should be carried out in ways consistent with rights norms and principles, including full and meaningful participation. Monitoring and evaluation should be ongoing. The assessment should be supported by outcome, process and structural indicators developed and used from the beginning of the process and further refned at each step. Where possible, the assessment should address root causes for the non-realisation of rights. These would include issues of empowerment and rights holders' and duty bearers' capacity. The assessment should be complementary to and integrated with other environmental and social impact assessments.

References

Beder, S. *Environmental principles and policies: An interdisciplinary introduction.* University of New South Wales Press and Earthscan, London, UK. 2006.

Dowie, M. Conservation refugees: When protecting nature means kicking people out. *Orion Online* November-December: 1–12. 2005.

International Council on Human Rights Policy (ICHRP). *Climate change and human rights: A rough guide.* ICHRP, Versoix, Switzerland. 2008.

Sensi, S. Human rights and the environment: A practical guide for environmental activists. *Policy Matters 15: Conservation and Human Rights.* CEESP/IUCN, CENESTA, Tehran, Iran. 2007.

United Nations Development Programme (UNDP). *Applying a human rights-based approach to development cooperation and programming: A UNDP capacity development resource.* 2006.

8

Conservation and Human Rights

In recent years concern has grown about the efects of conservation on local and indigenous communities. This concern comes both from within and beyond the conservation community. In particular, protected areas have come under close scrutiny; a long list of case studies highlights evictions, forced resettlement, and a reduction or loss of access to important resources and sources of income.

International law addresses the human rights violations resulting from evictions, but legal mechanisms relating to these laws are rarely accessible to individuals and communities in need of redress. This is not to say that international law is powerless or should be ignored. On the contrary, its power lies more in setting standards for state behaviour than in directly compelling states to protect or refrain from violating the human rights of their citizens. In addition to international law, many agencies that fund or implement conservation activities have adopted codes of practice, principles and internal policies to guide their activities and to minimise negative efects on local and indigenous communities.

It provides a brief overview of current provisions for addressing human rights in a conservation context—both 'hard' international law, and 'soft' law (guidelines, principles, etc.) developed by a range of organisations. It is intended as a reference to help quickly identify what is and is not currently addressed in international law and policy, what is binding and what is not, and what are practical requirements and what are aspirations.

The frst part of the sets out the most important 'hard' obligations concerning human rights and conservation to which states have committed themselves under international law—generally in the form of multilaterally agreed treaties. It should be noted, however, that within treaties, parties may agree to voluntary or indicative guidelines which are more akin to 'soft' law. The second part of the highlights the wide range of 'soft' law relevant to conservation and human rights. 'Soft' law is expressed in the resolutions, declarations, statements of principles, guidelines and action plans produced by UN agencies, international finance institutions and other multilateral organisations. More and more NGOs and other non-state actors also contribute to international 'soft' law both by participating in international fora and by issuing their own resolutions, declarations and statements of principles.

Hard Law Provisions

International law does not specifcally guarantee redress for people who have been displaced by conservation projects. Rather, obligations to these people are included more generally in international law on human rights, and on conservation and environmental protection.

International human rights law is basically concerned with imposing limitations on state sovereignty—it requires states to treat the populations within their jurisdiction according to internationally agreed standards. International environmental law recognises a state's sovereignty over its natural resources, subject to certain conditions. Provisions in 'hard' and soft' law relevant to the rights of local and indigenous communities and the role they play in conservation are set out below.

Multilateral Environmental Agreements (MEAs)

The last 35 years have seen signifcant growth in the number of multilateral environmental agreements (MEAs). Many MEAs deal with the conservation and sustainable use of natural and living resources.

Convention on Biological Diversity (CBD)

The *UN Convention on Biological Diversity* (CBD), agreed at the Earth Summit in Rio de Janeiro in 1992, is one of the most broadly subscribed international environmental treaties. The CBD has three main goals, the conservation of biodiversity, the sustainable use of its components and the equitable sharing of the benefts arising from the use of genetic resources. The CBD recognises the close and traditional dependence many indigenous and local communities have on biological resources (CBD, Preamble, para 12). Several provisions refer to this.

Article 8. In-situ conservation

Article 8 of the CBD not only recognises the 'interrelationship between the natural environment, sustainable development, and the well-being of indigenous peoples' but, under Article 8(j), Contracting Parties also specifcally commit themselves to respect, preserve and maintain the knowledge, innovations and practices of indigenous and local communities.

The Ad Hoc Open-ended Working Group on Article 8(j) produced a useful reference, a Compilation and Overview of Existing Instruments, Guidelines, Codes and Other Activities Relevant to the Programme of Work.

On the recommendation of the UN Permanent Forum on Indigenous Issues, the 7th Conference of the Contracting Parties (COP7) (CBD 2004) requested the Article 8(j) Working Group to develop an ethical code of conduct to ensure respect for the cultural and intellectual heritage of indigenous and local communities. This was reviewed at COP9 (CBD 2008) and includes:

1. *Intellectual Property:* Community and individual concerns over, and claims to, intellectual property relevant to traditional knowledge, innovations and practices related to the conservation and sustainable use of biodiversity should be acknowledged and addressed in the negotiation with traditional knowledge holders and/or indigenous and local communities, as appropriate, prior to starting activities/ interactions ... (G anx. sec 2(8))
2. *[Transparency/full disclosure]*: Indigenous and local communities should be [fully] informed [to the fullest extent possible] about the nature, scope and purpose of any proposed activities/interactions carried out by others [that may involve the use of their traditional knowledge, innovations and practices related to the conservation and sustainable use of biodiversity] ... (G anx. sec 2 (10))
3. *[Protection of] collective or individual ownership:* The resources and knowledge of indigenous and local communities can be collectively or individually owned. Those interacting with indigenous and local

communities should seek to understand the balance of collective and individual rights and obligations … (G anx. sec 2 (13))

4. *Fair and equitable sharing of benefts:* Indigenous and local communities ought to receive fair and equitable benefts for their contribution to any activities/interactions related to biodiversity and associated traditional knowledge [proposed to take place on, or which are likely to impact on, sacred sites and lands and waters traditionally occupied or used by indigenous and local communities] … (G anx. sec 2 (14))

5. *[Precautionary approach (including the concept of 'do no harm')]:* … the prediction and assessment of potential biological and cultural harms should include local criteria and indicators, and should fully involve the relevant indigenous and local communities. (G anx. sec 2 (16))

Article 10. Sustainable use of components of biological diversity

Article 10(c) requires Contracting Parties to protect and encourage customary uses of biological resources derived from traditional cultural practices that are compatible with the conservation of biological diversity and the sustainable use of its components.

Akwé: Kon Guidelines

The *Akwé: Kon Guidelines for the Conduct of Cultural, Environmental and Social Impact Assessments* regarding developments proposed to take place on, or which are likely to impact on, sacred sites and on lands and waters traditionally occupied or used by indigenous and local communities are an example of 'soft' law recommendations within the 'hard' law of a multilateral treaty. The guidelines were adopted in 2004 by COP7 as a collaborative framework for governments, indigenous and local communities, decision makers and development managers. While not binding, the COP requested governments to explore options for incorporating the guidelines into national legislation and policies. The guidelines require impact assessments to take the following into account (guideline 52):

(a) Prior informed consent of the afected indigenous and local communities;

(b) Gender considerations;

(c) Impact assessments and community development plans;

(d) Legal considerations;

(e) Ownership, protection and control of traditional knowledge, innovations and practices and technologies used in cultural, environmental and social impact assessment processes;

(f) Mitigation and threat-abatement measures;

(g) Need for transparency; and

(h) Establishment of review and dispute resolution procedures. (Secretariat of the Convention on Biological Diversity 2004)

Protected Areas Programme of Work

Protected areas are a key element of the CBD as well as being a key indicator for the achievement of Goal 7, on Environmental Sustainability, of the UN Millennium Development Goals (MDGs). COP7 decided that 'the establishment, management and monitoring of protected areas should take place with the full and efective participation, and the full respect for the rights of, indigenous and local communities consistent with domestic law and applicable international obligations'. In addition, COP7 adopted a programme of work on protected areas (PoW on PA). The PoW on PA highlights the close links between conservation and socio-economics and suggests that Parties:

2.1.1 Assess the economic and socio-cultural costs, benefts and impacts arising from the establishment and maintenance of protected areas, particularly for indigenous and local communities, and adjust policies to avoid and mitigate negative impacts, and where appropriate compensate costs and equitably share benefts in accordance with the national legislation …

2.2.4 Promote an enabling environment (legislation, policies, capacities and resources) for the involvement of indigenous and local communities and relevant stakeholders in decision making, and the development of their capacities and opportunities to establish and manage protected areas, including community-conserved and private protected areas.

2.2.5 Ensure that any resettlement of indigenous communities as a consequence of the establishment or management of protected areas will only take place with their prior informed consent that may be given according to national legislation and applicable international obligations.

(CBD PoW on PA, Programme Element 2: Governance, Equity, and Beneft Sharing)

Other MEAs

A number of other MEAs address the rights of indigenous and local communities in a conservation context. The *Convention on Wetlands* of International Importance especially as Waterfowl Habitat (Ramsar Convention), agreed in 1971, provides the framework for national action and international cooperation for the conservation and wise use of wetlands and their resources. In 1996 COP6 called upon Contracting Parties 'to make specifc eforts to encourage active and informed participation of local and indigenous people at Ramsar listed sites and other wetlands and their catchments, and their direct involvement, through appropriate mechanisms, in wetland management' These issues were addressed and elaborated on in COP7 with the adoption of Guidelines for establishing and strengthening local communities' and indigenous peoples' participation in the management of wetlands.

The *UN Convention to Combat Desertifcation* (UNCCD), while not always considered a conservation treaty, focuses on sustainable natural resource management as a mechanism to address land degradation. The emphasis is on 'bottom up' approaches that draw on local people and communities to develop and implement conservation programmes. This Convention also emphasises local participation in planning, designing and implementing conservation programmes (Article 3) and in National Action Programmes (Article 10).

The *Åarhus Convention* is the most far-reaching and detailed treaty to date on public participation in environmental decision making, bringing together human rights and environmental issues. Article 1 of the Convention recognises the right to a healthy environment for present and future generations and requests states to ensure the protection of this right through three fundamentals, access to information, public participation and access to justice.

International Human Rights Law

Numerous provisions in human rights law may be relevant to the treatment of indigenous and local communities afected by conservation practices. In this regard, it is worth remembering that state obligations to ensure human rights also have indirect efects on non-state actors. The most comprehensive catalogues of fundamental human rights are the *International Covenant on Civil and Political Rights* (ICCPR) and the *International Covenant on Economic, Social and Cultural Rights* (ICESCR). Both are legally binding treaties signed by most of the world's states. It shows the provisions most relevant to the treatment of local and indigenous communities.

Human Rights Law and Conservation Practices

Right to Life (Article 6 ICCPR)

The right to life not only prohibits the arbitrary or negligent taking of human life, but also sets out obligations. In the case of *Yanomani Indians v Brazil*, for example, the Inter-American Commission for Human Rights found that the construction of a Trans-Amazonian highway through the territory where the Indians lived impaired their traditional life style, amounting to a violation of their right to life.

The prohibition of arbitrary or unlawful interference with privacy, family, home or correspondence (Article 17 ICCPR)

This right aims to secure a sphere within which individuals may freely pursue the fulfilment and development of their private lives and physical wellbeing. The enjoyment of this right can be severely impaired by environmental conditions.

The right to freedom of movement and freedom to choose a place of residence (Article 12 ICCPR)

According to the Human Rights Committee, the right to liberty of movement and to reside in the place of one's choice includes protection against forced internal displacement and arbitrary denial of access to defned parts of a territory.

Prohibition of discrimination (Article 27 of the ICCPR and, more specifcally, Convention on the Elimination of All Forms of Racial Discrimination (CERD))

Persons belonging to ethnic, religious or linguistic minorities must not be denied the right to enjoy their own culture, profess and practice their own religion, or use their own language. States are required to put in place positive measures to prevent violations of these entitlements, both by state authorities and third parties. These rights are often associated with territory and the use of natural resources.

The right to adequate food and housing (Article 11 ICESCR)

The right to food includes feeding oneself directly from productive land or other natural resources. The right is inherently linked to social justiceb, requiring the adoption of appropriate economic, environmental and social policies oriented to the eradication of poverty. The right to adequate housing provides that all persons should possess a degree of security of tenure which guarantees legal protection against forced eviction, harassment and other threats. The ICESCR considers forced evictions prima facie incompatible with the Convention.

The right to self-determination (Article 1 of the ICCPR and the ICESCR)

A crucial aspect of the interpretation of the right to self-determination is the meaning of the term 'people'. According to the most widespread interpretation, 'people' applies to 'entire populations living in independent and sovereign states' and 'entire populations of territories that have yet to attain independence' (Cassese 1999, p. 59). It is unclear whether self-determination may be established without reference to a specifc territory. This is an important question for the applicability of this right to minorities.

The right to the highest attainable standard of health (Article 12 ICESCR)

The right to health relates to the enjoyment of a variety of facilities, goods, services and conditions necessary for the realisation of the highest attainable standard of health. This embraces a wide range of socio-economic factors that promote conditions in which people may lead a healthy life, such as adequate food and nutrition, housing, access to safe and potable water and adequate sanitation, safe and healthy working conditions and a healthy environment.

The right to development

The right to development was frst acknowledged in 1986 by the UN General Assembly, with the Declaration on the Right to Development. To date, however, the only binding human rights treaty including an explicit provision on the right to development is the African Charter on Human and Peoples' Rights.

The Rights of Indigenous People

The frst international law instrument covering indigenous rights was the 1959 International Labour Organization *Convention 107 Concerning the*

Protection and Integration of Indigenous and Other Tribal and Semi-Tribal Populations in Independent Countries. This Convention provides for a protection system to integrate and assimilate indigenous people in host states, an approach that was severely criticised for its lack of respect for indigenous identity. It has since been replaced by the 1989 International Labour Organization *Convention 169 Concerning Indigenous and Tribal Peoples in Independent Countries*. This seeks to preserve indigenous identity. Convention 169 emphasises indigenous people's relationships to territory, expressly recognising 'the rights of ownership and possession of the peoples concerned over the lands which they traditionally occupy' (ILO Convention 169, Article 14,1).

International Labour Organization Convention 169 articles

Article 14

1. The rights of ownership and possession of the peoples concerned over the lands which they traditionally occupy shall be recognised. In addition, measures shall be taken in appropriate cases to safeguard the right of the peoples concerned to use lands not exclusively occupied by them, but to which they have traditionally had access for their subsistence and traditional activities. Particular attention shall be paid to the situation of nomadic peoples and shifting cultivators in this respect.

Article 15

1. The rights of the peoples concerned to the natural resources pertaining to their lands shall be specially safeguarded. These rights include the right of these peoples to participate in the use, management and conservation of these resources.

Article 16

1. Subject to the following paragraphs of this Article, the peoples concerned shall not be removed from the lands which they occupy.
2. Where the relocation of these peoples is considered necessary as an exceptional measure, such relocation shall take place only with their free and informed consent. Where their consent cannot be obtained, such relocation shall take place only following appropriate procedures established by national laws and regulations, ... which provide the opportunity for efective representation of the peoples concerned.

3. Whenever possible, these peoples shall have the right to return to their traditional lands, as soon as the grounds for relocation cease to exist.
4. When such return is not possible ... these peoples shall be provided in all possible cases with lands of quality and legal status at least equal to that of the lands previously occupied by them, suitable to provide for their present needs and future development ...
5. Persons thus relocated shall be fully compensated for any resulting loss or injury.

Soft Laws

International 'soft' law is not binding *per se,* but plays a very important role in supporting international law. 'Soft' law sets the direction for formally binding obligations by informally establishing acceptable norms of behaviour and 'codifying' and refecting customary law. A number of 'soft' law provisions link human rights and conservation practices.

United Nations Agencies and Bodies

United Nations Department for Economic and Social Afairs

Agenda 21 is a major outcome of the 1992 UN Conference on Environment and Development, the Rio 'Earth Summit', and is the 'action plan' for implementing the Rio Principles. Agenda 21 relates environmental and natural resource management to human rights. For example, it states that 'people should be protected by law against unfair eviction from their homes or land'.

The *Johannesburg Plan of Implementation*, adopted at the World Summit on Sustainable Development (WSSD) in 2002, identifes the CBD as the key instrument for the conservation and sustainable use of biodiversity (WSSD Plan of Implementation, para 44). The plan includes actions to reduce the rate at which biodiversity is being lost:

(j) Subject to national legislation, recognize the rights of local and indigenous communities who are holders of traditional knowledge, innovations and practices ...

(l) Promote the efective participation of indigenous and local communities in decision and policy making concerning the use of their traditional knowledge.

United Nations Development Programme (UNDP)

Biodiversity is a focal area of the UNDP. UNDP is one of the implementing agencies of the Global Environment Facility (GEF) and an executor of the GEF Small Grants Fund. In 2001 the UNDP set out principles for relationships with indigenous people. The *UNDP and Indigenous Peoples: A Practice Note on Engagement* states:

27. By incorporating the 'right to development' in its work, UNDP fosters the full participation of indigenous peoples in its development processes and the incorporation of indigenous perspectives in development planning and decision-making …

29. … UNDP promotes the recognition of indigenous rights to lands, territories and resources; laws protecting indigenous lands; and the inclusion of indigenous peoples in key legislative processes.

United Nations Ofce of the High Commissioner of Human Rights (OHCHR)

The OHCHR Guiding Principles on Internal Displacement (1998) compile and restate human rights and humanitarian law relevant to internally displaced persons.

Principle 6/1: 1. Every human being shall have the right to be protected against being arbitrarily displaced from his or her home or place of habitual residence.

Principle 7/1: Prior to any decision requiring the displacement of persons, the authorities concerned shall ensure that all feasible alternatives are explored in order to avoid displacement altogether. Where no alternatives exist, all measures shall be taken to minimize displacement and its adverse efects.

Principle 7/3. If displacement occurs in situations other than during the emergency stages of armed conficts and disasters …

(a) Adequate measures shall be taken to guarantee those to be displaced full information on the reasons and procedures for their displacement and, where applicable, on compensation and relocation;

(b) The free and informed consent of those to be displaced shall be sought;

(c) The authorities concerned shall endeavour to involve those afected, particularly women, in the planning and management of their relocation;

(d) The right to an efective remedy, including the review of such decisions by appropriate judicial authorities, shall be respected.

Principle 9: States are under a particular obligation to protect against the displacement of indigenous peoples, minorities, peasants, pastoralists and other groups with a special dependency on and attachment to their lands.

The United Nations Declaration on the Rights of Indigenous Peoples (UN-DECRIPS) 2007

The *United Nations Declaration on the Rights of Indigenous Peoples* is non-binding but sets out a universal framework of rights that countries should recognise, guarantee and implement. UN-DECRIPS recognises indigenous people's right to own and control their lands and, to various degrees, their rights to own, use and manage the natural resources on those lands. The right to development is understood as the right to decide the *kind* of development that takes place on their lands and territories in accordance with their own priorities and cultures. States are called upon to consult with indigenous people and obtain their free and informed consent prior to approval of any project afecting their lands and resources.

Guidelines on Indigenous Peoples' Issues 2008

The Guidelines on Indigenous Peoples' Issues help UN Country Teams to integrate indigenous people's issues into country policies and programmes. Tey direct that 'Programming should encourage the development of human capabilities and the participation of indigenous peoples in community and social contexts, policy design and implementation at local, national, regional and global levels, creating strategies that can help them escape poverty'

Funding Agencies

World Bank

The World Bank *Operational Policies* (OPs) set standards and conditions that are binding for World Bank staf, grantees and borrowers.

Operational Policy on Involuntary Resettlement (OP 4.12), December 2001, Revised April 2004

(a) Involuntary resettlement should be avoided where feasible, or minimized, exploring all viable alternative project designs.

(b) Where it is not feasible to avoid resettlement, resettlement activities should be conceived and executed as sustainable development programs, providing sufcient investment resources to enable the persons displaced by the project to share in project benefts. Displaced persons should be meaningfully consulted and should have opportunities to participate in planning and implementing resettlement programs.

(c) Displaced persons should be assisted in their eforts to improve their livelihoods and standards of living, or at least to restore them, in real terms, to pre-displacement levels or to levels prevailing prior to the beginning of project implementation, whichever is higher.

This policy covers direct economic and social impacts that both result from Bank-assisted investment projects, and are caused by:

(a) the involuntary taking of land resulting in
   (i) relocation or loss of shelter;
   (ii) loss of assets or access to assets; or
   (iii) loss of income sources or means of livelihood, whether or not the afected persons must move to another location; or

(b) the involuntary restriction of access to legally designated parks and protected areas resulting in adverse impacts on the livelihoods of the displaced persons.

Operational Policy on Forests (OP 4.36), January 2002

10. To be acceptable to the Bank, a forest certifcation system must require:
    b) recognition of and respect for any legally documented or customary land tenure and use rights as well as the rights of indigenous peoples and workers;

11. In addition ... a forest certifcation system must ... be developed with the meaningful participation of local people and communities; indigenous peoples; non-governmental organizations representing consumer, producer, and conservation interests; and other members of civil society, including the private sector ....

Revised Operational Policy on Indigenous Peoples (OP 4.10), July 2005

1. This policy contributes to the Bank's mission of poverty reduction and sustainable development by ensuring that the development process fully respects the dignity, human rights, economies, and cultures of Indigenous Peoples. For all projects that are proposed for Bank financing and afect Indigenous Peoples, the Bank requires the borrower

to engage in a process of free, prior, and informed consultation. The Bank provides project financing only where free, prior, and informed consultation results in broad community support to the project by the afected Indigenous Peoples ....

Physical Relocation of Indigenous Peoples

20. Because physical relocation of Indigenous Peoples is particularly complex and may have signifcant adverse impacts on their identity, culture, and customary livelihoods, the Bank requires the borrower to explore alternative project designs to avoid physical relocation of Indigenous Peoples. In exceptional circumstances, when it is not feasible to avoid relocation, the borrower will not carry out such relocation without obtaining broad support for it from the afected Indigenous Peoples' communities as part of the free, prior, and informed consultation process.

In such cases, the borrower prepares a resettlement plan in accordance with the requirements of OP 4.12 ....

International Finance Corporation (IFC)

The International Finance Corporation is the private sector arm of the World Bank Group. World Bank Group policies and standards are binding on the IFC and its clients.

Policy on Social and Environmental Sustainability, 2006

8. Central to IFC's development mission are its eforts to carry out its investment operations and advisory services in a manner that 'do no harm' to people or the environment. Negative impacts should be avoided where possible, and if these impacts are unavoidable, they should be reduced, mitigated or compensated for appropriately.
19. ... IFC requires clients to engage with afected communities through disclosure of information, consultation, and informed participation, in a manner commensurate with the risks to and impacts on the afected communities.
20. IFC is committed to working with the private sector to put into practice processes of community engagement that ensure the free, prior, and informed consultation of the afected communities.

Performance Standards 2006

5: Land Acquisition and Involuntary Resettlement

- The client will consider feasible alternative project designs to avoid or at least minimize physical or economic displacement, while balancing environmental, social, and financial costs and benefts.
- When displacement cannot be avoided, the client will ofer displaced persons and communities compensation for loss of assets at full replacement cost and other assistance to help them improve or at least restore their standards of living or livelihoods ....
- Following disclosure of all relevant information, the client will consult with and facilitate the informed participation of afected persons and communities, including host communities, in decision-making processes related to resettlement ....
- The client will establish a grievance mechanism ... to receive and address specifc concerns about compensation and relocation that are raised by displaced persons or members of host communities

7: Indigenous Peoples: The client will establish an ongoing relationship with the afected communities of Indigenous Peoples from as early as possible in the project planning and throughout the life of the project. In projects with adverse impacts on afected communities of Indigenous Peoples,

Rights-based approaches Exploring issues and opportunities for conservation the consultation process will ensure their free, prior, and informed consultation and facilitate their informed participation on matters that afect them directly, such as proposed mitigation measures, the sharing of development benefts and opportunities, and implementation issues ....

Regional Development Banks

The policies relating to the involuntary resettlement of indigenous people of a number of regional development banks, including the Asian Development Bank (ADB), Inter-American Development Bank (IADB) and European Bank for Reconstruction and Development (EBRD), are similar to those of the World Bank and IFC. These policies are binding on bank staf, the governments of the borrowing countries and/or other private project sponsors and include:

Asian Development Bank (ADB) Policy on Involuntary Resettlement, 1995 (R1-79-95): As with the other IFIs [international financial institutions], the objectives of ABD's policy on involuntary resettlement are to avoid wherever possible and minimise the efects where it is unavoidable including (i) compensation for lost assets and loss of livelihood and income, (ii) assistance for relocation including provision of relocation sites with appropriate facilities and services, and (iii) assistance for rehabilitation to achieve at least the same level of well-being with the project as without it.

Asian Development Bank (ADB) Policy on Indigenous Peoples, 1998 This policy is intended to ensure that ADB interventions afecting indigenous people are consistent with their needs and aspirations; planned and implemented with their informed participation; equitable; and provide appropriate and acceptable compensation in case of negative efects.

Inter-American Development Bank (IADB) Strategies and Procedures on Socio-cultural Issues as Related to the Environment, 1995 These focus on the need to avoid resettlement where possible; the need for local consultation and participation in project design and implementation; and the need for capacity building in agricultural and other productive activities, where relocation can not be avoided 'so as to ensure the long-term economic viability of the new communities'.

Inter-American Development Bank (IADB) Operational Policy on Involuntary Resettlement (OP 710), 1998 This seeks to avoid or minimise the need for involuntary resettlement to the extent of reconsidering the project if large-scale relocation is inevitable. Where relocation does occur, the need for 'fair and adequate compensation and rehabilitation' is required. The policy includes special consideration of indigenous communities, emphasising that:

- customary rights will be fully recognized and fairly compensated;
- compensation options will include land-based resettlement; and
- the people afected have given their informed consent to the resettlement and compensation measures.

Involuntary Resettlement in Inter-American Development Bank (IADB) Projects. Principles and Guidelines, November 1999 These provide further guidance on the Operational Policy, including on compensation and arbitration procedures.

Inter-American Development Bank (IADB) Operational Policy on Indigenous Peoples (OP 765), February 2006 This includes specifc clauses on natural resource management and protected area projects including the following safeguards:

(i) respect for the rights recognized in accordance with the applicable legal norms;

(ii) in projects for natural resource extraction and management and protected areas management, the inclusion of:

   (1) prior consultation mechanisms to safeguard the physical, cultural, and economic integrity of the afected peoples and the sustainability of the protected areas and natural resources;

   (2) mechanisms for the participation of indigenous peoples in the utilization, administration, and conservation of these resources;

   (3) fair compensation for any damage these peoples might sufer as a result of the project; and

   (4) whenever possible, participation in project benefts.

European Bank for Reconstruction and Development (EBRD) Environmental and Social Policy, 2008 This includes specifc performance standards for projects involving indigenous people and requires projects to 'avoid adverse impacts of projects on the lives and livelihoods of Indigenous Peoples' communities, or when avoidance is not feasible, to minimise, mitigate, or compensate for such impacts.'

Global Environment Facility

The Global Environmental Facility (GEF) is the designated financial mechanism for a number of multilateral environmental agreements (MEAs). It has no specifc social or environmental policies, although a report by Grifths for the Forest People's Programme notes that 'although the GEF has, since 1994, adopted a generally understood rule that it does not fund involuntary resettlement, this crucial institutional safeguard has yet to be consolidated in official GEF policies'. Nevertheless, GEF does have a policy on Public Involvement and this notes that... 'all public involvement activities should be based on local needs and conditions... biodiversity projects afecting indigenous communities may require more extensive stakeholder participation than global projects which focus on technical assistance and capacity building at the national and regional levels' Guidelines on Public Involvement in Projects Financed by the GEF.

Bilateral aid Agencies

The Development Assistance Committee of the Organisation for Economic Cooperation and Development (DAC-OECD) is made up of the official development assistance agencies of OECD countries. DAC-OECD has developed a series of non-binding *G*uidelines on Aid and Environment, including the 1992 Guidelines for Aid Agencies on Involuntary Displacement and Resettlement in Development Projects. These call on members to avoid or minimise involuntary displacement wherever possible and state that 'Donor countries should not support projects that cause population displacement unless they contain acceptable resettlement plans protecting the rights of afected groups.' In particular, 'Indigenous groups, ethnic minorities, and pastoralists who may have informal customary rights to the land or other resources taken for the project must be provided with adequate land, infrastructure, and other compensation. The absence of legal title to land by such groups should not be a bar to compensation'.

International Conservation NGOs

Conservation International (CI)

In 2003, Conservation International created the Indigenous and Traditional Peoples Initiative. This initiative is designed to ensure and support the development of appropriate tools, knowledge and resources to enable traditional groups to continue efcient and efective stewardship of their land, and achieve sustainable community development.

Indigenous Peoples and Conservation International: Principles for Partnerships, 2003

3. … We will openly inform, consult and obtain the informed consent of formal representatives of indigenous groups prior to undertaking any actions that are directly tied to indigenous peoples, their territories or natural resources.
4. … We support eforts by indigenous groups to gain legal designation and management authority over ancestral lands and their resources, while respecting issues of national sovereignty.

8. … Our support includes enhancing the capacity of indigenous people's organizations and communities to prepare, implement, monitor and evaluate conservation activities or activities that are likely to have an impact upon conservation.
9. We recognize that there are often overlaps between lands set aside for legally designated parks and protected areas and lands customarily owned or used by indigenous peoples. CI recognizes both the signifcance of these customary rights and the need for long-term sustainable management of critical ecosystems. In legally designated parks and protected areas, CI will work with protected area and indigenous authorities to support collaborative management initiatives that recognize customary uses while ensuring that natural resources are not depleted and that actively involve indigenous communities in planning, zoning, and monitoring

International Union for Conservation of Nature (IUCN)

The International Union for Conservation of Nature has produced a raft of resolutions and recommendations on conservation and human rights at its World Congresses (previously General Assemblies) and other events, such as the once-a-decade World Parks Congress. These guide the work of IUCN, but are not binding on members. The most pertinent resolutions are:

- 12th IUCN General Assembly. *Resolution 12.5 Protection of Traditional Ways of Life* calls on governments to recognise indigenous people's rights to land particularly in the context of preventing displacement in conservation areas.
- 19th IUCN General Assembly. *Resolution 19.22 Indigenous People* urges governments to guarantee respect for the rights of local and indigenous people in protected areas.
- 1996 World Conservation Congress (Montreal) The 1996 IUCN World Conservation Congress adopted *Resolution WCC 1.53 Indigenous Peoples and Protected Areas,* which stresses the need to recognise the rights of indigenous people with regard to their lands and territories that fall within protected areas.

In 1999, IUCN, the World Commission on Protected Areas (WCPA) and the World Wide Fund for Nature (WWF) developed a set of Principles and Guidelines on Indigenous and Traditional Peoples and Protected Areas in response to Resolution WCC 1.53. These include guidance on protected area management agreements, noting that they should not only be based on respect for indigenous and traditional rights, but should also highlight the conservation responsibilities of indigenous peopl.

2000 World Conservation Congress—*IUCN Policy on Social Equity in Conservation and Sustainable use of Natural Resources*. Building on previous resolutions and indigenous rights conventions this policy notes that

> IUCN aims to: Respect indigenous people's knowledge and innovations, and their social, cultural, religious and spiritual values and practices. Recognise the social, economic and cultural rights of indigenous peoples such as their right to lands and territories and natural resources, respecting their social and cultural identity, their customs, traditions and institutions. Ensure full and just participation of indigenous peoples in all conservation activities supported and implemented by IUCN. Support indigenous peoples' right to make their own decisions afecting their lands, territories and resources, by assuring their rights to manage natural resources, such as wildlife, on which their livelihoods and ways of life depend, provided they make sustainable use of natural resources.

5th World Parks Congress (2003) The *Durban Accord* urges commitment to '... ensuring that people who beneft from, or are afected by protected areas have the opportunity to participate in relevant decision making on a fair and equitable basis in full respect of their human and social rights'. The *Durban Action Plan*, the plan to implement the Accord, includes a number of human rights targets:

Key Target 8: all existing and future protected areas shall be managed and established in full compliance with the rights of indigenous peoples, mobile peoples and local communities.

Key Target 9: protected areas shall have representatives chosen by indigenous peoples and local communities in their management proportionate to their rights and interests.

Key Target 10: participatory mechanisms for the restitution of indigenous peoples' traditional lands and territories that were incorporated in protected areas without their free and informed consent established and implemented by 2010.

In addition to the Durban Accord and Durban Action Plan, the 5th World Parks Congress generated a number of more detailed recommendations including:

WPC Recommendation 24: Indigenous Peoples and Protected Areas which recommends that existing and future protected areas respect the rights of indigenous people through cessation of all involuntary evictions; full, prior, informed consent in establishing protected areas; and establishing compensation and restitution mechanisms to address historical injustices.

WPC Recommendation 27: Mobile Indigenous Peoples and Conservation which is the frst instrument to focus specifcally on mobile communities and call for recognition of their resource management systems.

- 2004 World Conservation Congress (Bangkok) *Resolution 3.015 Conserving nature and reducing poverty by linking human rights and the environment* encourages IUCN to include human rights in its mission and calls on the IUCN Commission on Environmental Law to, among other things, analyse human rights law to provide efective access to justice in the event of the violation of rights and to 'provide a progress report to future World Conservation Congresses summarizing legal developments in human rights law and litigation that are pertinent to IUCN's Mission, with an emphasis on human-rights tools that may be used by IUCN and its members in pursuit of the Mission'.
- 2008 World Conservation Congress (Barcelona) passed a number of resolutions on human rights. In particular:

Resolution 4.056 Rights-based approaches to conservation which calls on IUCN members and other actors to:

(a) develop and/or work towards application of rights-based approaches, to ensure respect for, and where possible further fulflment of human rights, tenure and resource access rights, and/or customary rights of indigenous peoples and local communities in conservation policies, programmes, projects, and related activities;

(b) encourage relevant government agencies, private actors, businesses and civil-society actors to monitor the impacts of conservation activities on human rights as part of a rights based approach;

(c) encourage and establish mechanisms to ensure that private-sector entities fully respect all human rights, including Indigenous Peoples' rights, and take due responsibilities for the environmental and social damage they engender in their activities; and

(d) promote an understanding of responsibilities and synergies between human rights and conservation.

Resolution 4.052 Implementing the United Nations Declaration on the Rights of Indigenous Peoples which endorses the UN Declaration, and calls on all IUCN members to do likewise and to integrate it into their work programmes.

Resolution 4.048 Indigenous Peoples, protected areas and implementation of the Durban Accord which reinforces this request, especially with respect to protected areas.

Resolution 4.053 Mobile Indigenous Peoples and biodiversity conservation which calls on IUCN to adhere to the fve principles of the *Dana Declaration* 2002, the outcome of a meeting of social and natural scientists, and NGOs. These principles include the principle that 'conservation approaches with potential impact on mobile peoples and their natural resources must recognise mobile peoples' rights ....

Wildlife Conservation Society (WCS)

The Wildlife Conservation Society has a policy on human displacement. This recognises that WCS, as a part of its work, is often required to advise authorities on access to local resources. The policy notes that WCS only rarely and as a last resort advises authorities on the displacement of people from particularly fragile, valuable, or dangerous environments.

WCS Policy on Human Displacement and Modifcation of Resource Access to Achieve Conservation Objectives 2007

This policy states that advice provided to authorities by WCS will take into account the legitimacy of land and resource claims, and the vulnerability of the people afected. Where the decision is to displace people or restrict access to resources the WCS policy is to make every efort to ensure that:

2.1. The authorities obtain in advance the freely given and informed consent of all persons proposed to be displaced or to lose resource access;

2.2. The authorities seek to minimize the impact on the people proposed to be displaced or to lose resource access;

2.3. The authorities take into account both the material and nonmaterial needs of the people proposed to be displaced or lose resource access and seek to provide them with reasonably acceptable resettlement alternatives or opportunities to secure comparable or enhanced means of livelihood;

2.4. The authorities meet all their legal and contractual obligations to the people ....

World Wide Fund for Nature/World Wildlife Fund (WWF)

In 1996, WWF issued a Statement of Principles on Indigenous Peoples and Conservation. This statement provides guidance on partnering with indigenous people's organisations to conserve biodiversity within indigenous lands and territories, and on promoting the sustainable use of natural resources. The principles were updated in 2008.

Indigenous Peoples and Conservation: WWF Statement of Principles, 2008

Since indigenous peoples are often discriminated against and politically marginalized, WWF is committed to make special eforts to respect, protect, and comply with their collective and individual rights, including customary as well as resource rights, in the context of conservation initiatives. This includes, but is not limited to, those set out in national and international law, and in other international instruments. In particular, WWF fully endorses the provisions about indigenous peoples contained in the following international instruments: Agenda 21; Convention on Biological Diversity; ILO Convention 169 (Convention Concerning Indigenous and Tribal Peoples in Independent Countries); UN Declaration on the Rights of Indigenous Peoples.

The principles specify some of the rights that WWF recognises and strives to uphold, many of which are mentioned in the documents described above. These include the rights of indigenous peoples to land and resources, and the right to determine development pathways. WWF also recognises that claims to land and resources are often contested and that the rights of indigenous peoples take priority: 'In instances where multiple local groups claim rights to resources in indigenous territories, WWF recognizes the primary rights of indigenous peoples ... due regard for the rights and welfare of other legitimate stakeholders ....'

In addition to recognising the rights of indigenous peoples, the principles also identify WWF responsibilities in engaging with indigenous people. These include encouraging governments and other stakeholders to recognise indigenous land rights; ensuring WWF practises due diligence in exploring historic and current land claims and rights before initiating any conservation activities; and working with indigenous groups to challenge any activities which are proceeding without prior, informed consen.

Other Conservation Organisations

The conservation organisations mentioned above have—publicly or internally—codifed their policies and approaches regarding indigenous and local community rights. Other conservation organisations have made general statements on their approaches to human rights issues. The Nature Conservancy (TNC), for example, has a set of *Core Values* including a *Commitment to People* which notes, 'We respect the needs of local communities by developing ways to conserve biological diversity while enabling them to live productively and sustainably' Similarly Fauna and Flora International (FFI) pledges that it will '... take account of human needs'while BirdLife International aims to 'help, through birds, to conserve biodiversity and to improve the quality of people's lives'.

Conservation organisations also make joint statements on indigenous and local community rights. During the UN World Summit in 2005, for example, Peter Seligmann, Chairman and CEO of Conservation International, announced on behalf of BirdLife International, Conservation International, Fauna and Flora International, Wildlife Conservation Society, The Nature Conservancy and the World Wide Fund for Nature that 'A group of the world's largest conservation and environmental NGOs have come together to announce tonight their commitment to integrate conservation and development eforts. After working for decades in some of the world's most impoverished places, the environmental community knows the critical need to work for improved livelihoods, respect indigenous and vulnerable peoples, and seek sustainable responses to the root causes that lead jointly to poverty and ecological disruption.'

Indigenous Organisations

Just as many of the conservation agencies have developed policies, principles and statements on indigenous and local community rights, indigenous organisations have commented publicly on the responsibilities of conservation organisations.

Coordinating Body for the Indigenous Organizations of the Amazon Basin (COICA)

In the late 1980s the Coordinating Body for the Indigenous Organizations of the Amazon Basin, produced a key text on indigenous rights and the links to both conservation and development. *T*wo Agendas on Amazon Development addressed both conservation and development. Part One, directed at the development community, called for recognition of indigenous people's rights and the need for prior, informed consent in any development intervention. Part Two, To the Community of Concerned Environmentalists, acknowledged the role of the international conservation community in rainforest conservation but expressed concern about the preoccupation with wildlife over and above the needs of local communities. As a follow up, the First Summit between Indigenous Peoples and Environmentalists was held in Iquitos, Peru, in 1990. The resulting *Iquitos Declaration* confrmed the importance of recognising indigenous land rights and led to the establishment, in 1993, of the Coalition in Support of Amazonian Peoples and Teir Environment

International Alliance of Indigenous and Tribal Peoples of the Tropical Forests

The International Alliance of Indigenous and Tribal Peoples of the Tropical Forests (IAITPTF), founded in 1992, has members from all over the tropics. In 2002, the Alliance adopted a charter that includes articles relating to biodiversity and conservation:

- Article 42. Conservation programmes must respect our rights to the use and ownership of the territories and resources we depend on. No programmes to conserve biodiversity should be promoted on our territories without our free, prior and informed consent as expressed through our indigenous organizations.

References

Barutciski, M. International law and development-induced displacement and resettlement. *In:* De Wet, C.J. (ed.) *Development-induced displacement: Problems, policies and people.* Berghan Books, Oxford, UK and New York, NY, USA. 2006.

Birnie, P.W. and Boyle, A. *International law and the environment.* Oxford University Press. Oxford, UK and New York, NY, USA.2006.

Cassese, A. Self-determination of peoples: *A legal reappraisal* (Hersch Lauterpacht Memorial Lectures), Cambridge University Press, Cambridge, UK.1999.

Driesen, D.M. 'Thirty years of international environmental law: A retrospective and plea for reinvigoration', 30 Syracuse J. Int'l L. and Com. 353. 2003.

9

World Conservation Strategy

The World Conservation Strategy (WCS) was commissioned by the United Nations Environment Programme (UNEP) which together with the World Wildlife Fund (WWF) provided the financial support for its preparation and contributed to the evolution of its basic themes and structure. IUCN is grateful to both organizations for all their support. While the WCS has been prepared by IUCN and primarily reflects IUCN's views and approaches it is intended that the Strategy represent a consensus of policy on conservation efforts in the context of world development. To this end the final draft was submitted to the Food and Agriculture Organization of the United Nations (FAO) and the United Nations Educational, Scientific and Cultural Organization (Unesco), as well as to UNEP and WWF, and all four organizations carefully reviewed it and made significant contributions to it. The WCS is endorsed by the Ecosystem Conservation Group (ECG) the members of which are UNEP, FAO, Unesco and'IUCN.

The WCS, as the product of an extremely thorough consultation process, inevitably reflects a compromise: among conservationists, who may differ on the relative importance of particular ecosystems, species, issues and measures; and between conservationists and the practitioners of development, who may differ in their emphasis on maintenance on the one hand and production on the other. Knowledge of the world is so patchy that global generalizations are particularly prone to error. These drawbacks are recognized; but they are considered less important than the need to present a statement of agreed conservation requirements and priorities, around which conservationists and development practitioners alike could rally, and to adopt a perspective unconfined by the boundaries that separate but do not insulate nation from nation, sector from sector, or interest from interest.

There is a parallel, paperback version of the World Conservation Strategy for the general reader. The paperback differs from the version in this pack in style and layout and in providing both a fuller account of the importance of living resource conservation and a more detailed description of the priority conservation issues. The paperback devotes less space than the pack to the organizational aspects of conservation and its integration with development. Its purpose is to make more people aware of the vital importance of living resource conservation.

As a supplement to the two versions of the WCS, a Sourcebook will be published over several years in a number of volumes—on species, on ecosystems of the land, fresh waters and the sea, and on issues and measures. The Sourcebook will provide more detailed information than can be contained in the Strategy, and a basis for the Strategy's further development. The WCS is intended to be an evolving effort and is expected to be up-dated and improved from time to time, in response to new knowledge, better understanding, changes in perception and values, and conservation progress as a result of the Strategy's implementation. For it is strongly hoped that governments, nongovernmental organizations and intergovernmental bodies will be quick to carry out the relevant recommendations. The members of the Ecosystem Conservation Group and WWF are ready to give whatever help they can within the limits of available resources, should their assistance be requested.

The aim of the World Conservation Strategy is to help advance the achievement of sustainable development through the conservation of living resources. The Strategy:

1. explains the contribution of living resource conservation to human survival and to sustainable development;
2. identifies the priority conservation issues and the main requirements for dealing with them;
3. proposes effective ways for achieving the Strategy's aim.

The Strategy is intended to stimulate a more focussed approach to living resource conservation and to provide policy guidance on how this can be carried out. It concentrates on the main problems directly affecting the achievement of conservation's objectives; and on how to deal with them through conservation. In particular, the Strategy identifies the action needed both to improve conservation efficiency and to integrate conservation and development.

The Strategy is intended chiefly for three groups of user (none of which is wholly separate from the others):

1. Government policy makers and.their advisers. Few governments have the financial and technical resources to address all of the problems of living resource conservation at once. Therefore they need to know what needs to be done first. Accordingly, the Strategy both recommends ways of overcoming the main obstacles to conservation and provides guidance on what action is most important. The Strategy is relevant to any level of government with significant responsibilities for planning and managing the use of living resources.
2. Conservationists and others directly concerned with living resources. For this group, the Strategy indicates those areas where conservation action is most urgently needed and where it is likely to yield the greatest and most lasting results. It also proposes ways in which conservation can participate more effectively in the development process, thereby increasing the likelihood of its being positively received by the development community and of helping to ensure that development is sustainable.
3. Development practitioners, including aid agencies, industry and commerce, and trade unions. For this group the Strategy demonstrates that conservation improves the prospects of sustainable development and proposes ways of integrating conservation into the development process. It also attempts to identify those areas where the interests of conservation and of development are most likely to coincide and therefore where a closer partnership between the two processes would be particularly advantageous to both.

Living Resource Conservation for Sustainable Development

Earth is the only place in the universe known to sustain life. Yet human activities are progressively reducing the planet's life-supporting capacity at a time when rising human numbers and consumption are making increasingly heavy demands on it. The combined destructive impacts of a poor majority struggling to stay alive and an affluent minority consuming most of the world's resources are undermining the very means by which all people can survive and flourish.

Humanity's relationship with the biosphere (the thin covering of the planet that contains and sustains life) will continue to deteriorate until a new international economic order is achieved, a new environmental ethic adopted, human populations stabilize, and sustainable modes of development become the rule rather than the exception. Among the prerequisites for sustainable development is the conservation of living resources.

Development is defined here as: the modification of the biosphere and the application of human, financial, living and non-living resources to satisfy human needs and improve the quality of human life. For development to be sustainable it must take account of social and ecological factors, as well as economic ones; of the living and non-living resource base; and of the long term as well as the short term advantages and disadvantages of alternative actions.

Conservation is defined here as: the management of human use of the biosphere so that it may yield the greatest sustainable benefit to present generations while maintaining its potential to meet the needs and aspirations of future generations. Thus conservation is positive, embracing preservation, maintenance, sustainable utilization, restoration, and enhancement of the natural environment. Living resource conservation is specifically concerned with plants, animals and microorganisms, and with those non-living elements of the environment on which they depend. Living resources have two important properties the combination of which distinguishes them from non-living resources: they are renewable if conserved; and they are destructible if not.

Conservation, like development, is for people; while development aims to achieve human goals largely through use of the biosphere, conservation aims to achieve them by ensuring that such use can continue. Conservation's concern for maintenance and sustainabihty is a rational response to the nature of living resources (renewability + destructibility) and also an ethical imperative, expressed in the belief that "we have not inherited the earth from our parents, we have borrowed it from our children".

Conservation is a process—to be applied cross-sectorally—not an activity sector in its own right. In the case of sectors (such as agriculture, fisheries, forestry and wildlife) directly responsible for the management of living resources, conservation is that aspect of management which ensures that utilization is sustainable and which safeguards the ecological processes and genetic diversity essential for the maintenance of the resources concerned. In the case of other sectors (such as health, energy, industry), conservation is that aspect of management which ensures that the fullest sustainable advantage is derived from the living resource base and that activities are so located and conducted that the resource base is maintained.

Living resource conservation has three specific objectives:

a. *to maintain essential ecological processes and life-support systems* (such as soil regeneration and protection, the recycling of nutrients, and the cleansing of waters), on which human survival and development depend;

b. *to preserve genetic diversity* (the range of genetic material found in the world's organisms), on which depend the breeding programmes necessary for the protection and improvement of cultivated plants and domesticated animals, as well as much scientific advance, technical innovation, and the security of the many industries that use living resources;

c. *to ensure the sustainable utilization of species and ecosystems* (notably fish and other wildlife, forests and grazing lands), which support millions of rural communities as well as major industries[1].

Living resource conservation is just one of a number of conditions necessary to assure human survival and wellbeing, and a world conservation strategy is but one of a number of strategies needed: a strategy for peace; a strategy for a new international economic order; a strategy for human rights; a strategy for overcoming poverty; a world food supply strategy; a population strategy. Several of these issues are properly the subject of the International Development Strategy for the Third United Nations Development Decade. All such strategies should be mutually reinforcing. None has much chance of success unless they are. The integration of conservation and development is particularly important, because unless patterns of development that also conserve living resources are widely adopted, it will become impossible to meet the needs of today without foreclosing the achievement of tomorrow's.

Conservation and development have so seldom been combined that they often appear—and are sometimes represented as being—incompatible. Conservationists themselves have helped— quite unwittingly—to foster this misconception. Too often they have allowed themselves to be seen as resisting all development—although often they have been forced into that posture because they have not been invited to participate in the development process early enough. The result has been not to stop development, but to persuade many development practitioners, especially in developing countries, that conservation is not merely irrelevant, it is harmful and anti-social. Consequently, development has continued unimpeded by conservationists yet with the seeds of its eventual failure lying in the ecological damage that conservation could have helped prevent.

That conservation and sustainable development are mutually dependent can be illustrated by the plight of the rural poor. The dependence of rural communities on living resources is direct and immediate. For the 500 million people who are malnourished, or the 1500 million people whose only fuel is wood, dung or crop wastes, or the almost 800 million people with incomes of $50 or less a year—for all these people conservation is the only thing between hem and at best abject misery, at worst death. Unhappily, people on the margins of survival are compelled by their poverty—and their consequent vulnerability to inflation—to destroy the few resources available to them. In widening circles around their villages they strip trees and shrubs for fuel until the plants wither away and the villagers are forced to burn dung and stubble. The 400 million tonnes of dung and crop wastes that rural people burn annually are badly needed to regenerate soils already highly vulnerable to erosion now that the plants that bind them are disappearing.

It would be wrong, however, to conclude that conservation is a sufficient response to such problems. People whose very survival is precarious and whose prospects of even temporary prosperity are bleak cannot be expected to respond sympathetically to calls to subordinate their acute short term needs to the possibility of long term returns. Conservation must therefore be combined with measures to meet short term economic needs. The vicious circle by which poverty causes ecological degradation which in turn leads to more poverty can be broken only by development. But if it is not to be self-defeating, it must be development that is sustainable—and conservation helps to make it so. The development efforts of many developing countries are being slowed or compromised by lack of conservation. In Southeast Asia excessive clearing of forests has caused fluctuations in river flow that are lowering rice yields. Throughout the developing world the lifetimes of hydroelectric power stations and water-supply systems are being cut as reservoirs silt up—because siltation is accelerated by deforestation, overgrazing and other unwise land uses.

The activities of every organism modify its environment, and those of human beings are no exception. Although environmental modification is both natural and a necessary part of development, this does not mean that all modification leads to development (nor that preservation impedes it). While it is inevitable that most of the planet will be modified by people and that much of it will be transformed, it is not at all inevitable that such alterations will achieve the social and economic objectives of development. Unless it is guided by ecological, as well as by other environmental, and by social, cultural and ethical considerations, much development will continue to have undesired effects, to provide reduced benefits or even to fail altogether. There is a close relationship between failure to achieve the objectives of con servation and failure to achieve the social and economic objectives of development—or, having achieved them, to sustain that achievement. Hence the goal of the World Conservation Strategy is the integration of conservation and development to ensure that modifications to the planet do indeed secure the survival and wellbeing of all people.

Objectives of Conservation and Requirements for Their Achievement

Maintenance of Essential Ecological Proce

Essential ecological processes are those processes that are governed, supported or strongly moderated by ecosystems and are essential for food production, health and other aspects of human survival and sustainable development. "Life-support systems" is shorthand for the main ecosystems involved—for example, watershed forests or coastal wetlands. The maintenance of such processes and systems is vital for all societies regardless of their stage of development. Many archaeological relics, whether of great civilizations or peasant villages, testify to the consequences of not doing so. Today, the most important and most threatened life-support systems are agricultural systems, forests, and coastal and freshwater systems.

Agricultural Systems

Only about 11 % of the world's land area (excluding Antarctica) offers no serious limitation to agriculture; the rest suffers from drought, mineral stress (nutritional deficiencies or toxicities), shallow depth, excess water, or permafrost. The best land is not evenly distributed. The world's cropland currently occupies 14 million km, and although it may be possible to double this area, much of the best land is already being farmed. Unfortunately large areas of prime quality land are being permanently taken out of agricultural use by being built on. In developed countries at least 3,000 km of prime agricultural land are submerged every year under urban sprawl: between 1960 and 1970 Japan lost 7.3% of its agricultural land to buildings and roads and European countries lost from 1.5% (Norway) to 4.3% (Netherlands). In addition, close to one third of the world's arable land will be destroyed in the next 20 years if current rates of land degradation continue.

Soil is a crucial life-support system, since the bulk of all food production depends on it. Soil erosion is a natural and continuous process, but in undisturbed ecosystems with a protective cover of plants the soil is usually regenerated at the same rate it is removed. If soil and vegetation are not in balance, as often they are not when influenced by poorly managed human activities, erosion is accelerated with disastrous consequences. Even under natural conditions of vegetation cover, nature takes from 100 to 400 years or more to generate 10 millimetres of top soil; and 3,000 to 12,000 years would be needed to generate soil to a depth of the length of this page. So once the soil has gone, for all practical purposes it has gone for good.

Soil loss has accelerated sharply throughout the food-hungry tropics, which are generally more susceptible to erosion than the temperate zone, due to the topography of the land and the nature of the soils and rainfall. More than half of India, for example, suffers from some.form of soil degradation: out of her total of 3.3 million km, 1.4 million km are subject to increased soil loss, while an additional 270,000 km are being degraded by floods, salinity and alkalinity. An estimated 6,000 million tonnes of soil are lost every year from 800,000 km alone; with them go more than 6 million tonnes of nutrients—more than the amount that is applied in the form of fertilizers.

The productivity of agricultural ecosystems depends not only on maintaining soil quality but also on retaining the habitats of beneficial insects and other animals, such as crop pollinators and the predators and parasites of pests. Effective pest control is no longer a matter of heavy applications of pesticides, partly because of the rising cost of petroleum-derived products but largely because excessive pesticide use promotes resistance, destroys natural enemies, turns formerly innocuous species into pests, harms other non-target species, and contaminates food and feed. Instead pesticides should be used to supplement a battery of methods integrated in appropriate combinations: these methods include introduction of pest-resistant crop varieties, special planting combinations and patterns, mechanical methods, the use of repellents and hormones, and encouragement of natural enemies.

Forests

Besides supplying timber and other products, forests have a vital effect on processes of great significance for people. They influence local and regional climates, generally by making them milder, and they help to ensure a continuous flow of clean water. Some forests, notably tropical cloud forests, even increase the availability of water by intercepting moisture from clouds. Watershed forests are particularly important because they protect soil cover on site and protect areas downstream from excessive floods and other harmful fluctuations in streamflow. By thus reducing the silt load of rivers, watershed forests also help prevent the clogging of reservoirs, irrigation systems, canals and docks, and the smothering by sediment of coral reefs.

Yet watershed forests are being widely devastated—by clearance for agriculture, by logging and cutting for fuel, by overgrazing, and by badly managed road building. The results can be extremely expensive. It costs Argentina $10 million a year to dredge silt from the estuary of the River Plate and keep Buenos Aires open to shipping: yet 80% of the 100 million tonnes of sediment that every year threaten to block the harbour comes from only 4% of the drainage basin—the heavily overgrazed catchment area of the Bermejo River 1,800 km upstream. In India the annual cost of damage by floods ranges from $140 million to $750 million..

Sedimentation as a result of careless use of watershed forests can cut drastically the economic life of reservoirs, hydroelectric facilities and irrigation systems. The capacity of India's Nizam-sagar reservoir has been more than halved (from almost 900 million m to fewer than 340 million m) and there is now not enough water to irrigate the 1,100 km of sugarcane and rice for which it was intended—and hence not enough sugarcane to supply local sugar factories. Deforestation in northern Luzon in the Philippines has silted up the reservoir of the Ambuklao Dam so fast that its useful life has been reduced from 60 to 32 years. Such problems are not confined to developing countries : for example, it has been estimated that more than 1,000 million m of sediment are deposited every year in the major reservoirs of the USA. Although they have not been calculated (indeed, probably cannot be), the global costs of sediment removal, river dredging, reconstruction of irrigation systems and loss of investment in expensive structures like dams must be huge. Only 10% of the world's population live in mountainous areas, but another 40% live in the adjacent plains; so the lives and livelihoods of half the world directly depend on the way in which watershed ecosystems are managed.

In areas under shifting cultivation forests also act to restore soil fertility. More than 200 million people occupying about 30 million km of tropical forests live by practising shifting cultivation—cropping an area for a few years, then clearing another area, leaving the first one fallow to revert to scrub and forest. The fallow period lasts from 8-12 years in tropical rain forests to 20-30 years in drier areas, and during this time the forest cover enables the soil to regenerate. This is a stable, productive practice if the population itself is stable; but if populations are growing, which nowa days they usually are, the pressure on land increases, fallow periods shorten, the soil has no chance to regenerate, and wider and wider tracts of otherwise productive forest land are destroyed. Almost two-thirds of the land under shifting cultivation is upland forest, much of it on steep slopes, and the resulting erosion is severe. In the Ivory Coast, shifting cultivation reduced the forest cover by 30% between 1956 and 1966 and now only 50,000 km remain out of the 150,000 km that is believed to have existed at the beginning of this century. Similarly, shifting cultivators clear about 3,500 km a year in the Philippines—in Mindanao alone they cleared 10,000 km between 1960 and 1971 n.

Coastal and Freshwater Systems

Coastal wetlands and shallows—especially estuaries and mangrove swamps—provide food and shelter for waterfowl and for fishes, crustaceans and molluscs utilized by an estimated two-thirds of the world's fisheries. Some are among the world's most lucrative fisheries, notably those for shrimp. Seagrass meadows also act as nurseries and nutrient suppliers for economically important fish species. Coral ecosystems are of more local, but nonetheless vital, significance—providing habitats for the fish on which many coastal communities in developing countries depend. In addition, coastal wetlands and coral reefs are extremely important for coastal protection : without the activities of corals and other reef-building organisms, for example, more than 400 islands would not exist. Similarly, many freshwater wetlands and floodplains support important inland fisheries, while floodplain agriculture has long relied on the regular supply of nutrients by floodwaters.

Wetlands, floodplains, seagrass beds and coral reefs are being destroyed the world over, with severe effects on the economies that depend on them most closely. For example, the cost of damage to US marine fisheries caused by degradation of coastal wetlands has been estimated to be almost $86 million a year. In Sri Lanka repeated removal of corals for the production of lime is so extensive that a local fishery has collapsed; mangroves, small lagoons and coconut groves have disappeared; and local wells have been contaminated with salt. In many parts of the world the construction of dams has blocked the passage of migrating fish and drowned or otherwise destroyed the habitats of others; and although the new reservoir may support a new fishery this does not always compensate for the loss of the floodplain. The habitats of many other aquatic animals also are threatened. For example, many sea turtle nesting beaches have been wrecked for roads, housing and other developments; while the breeding areas of gray whales and belugas (or white whales) are also at risk.

These are typical effects of the impacts on coastal and freshwater ecosystems everywhere: impacts such as industrial and agricultural pollution; the construction of dams; siltation from eroded uplands; filling to provide sites for industry, housing, recreation, airports and farmland; dredging to create, deepen or improve harbours; quarrying; and cutting of mangroves for fuel. As the commercially valuable fisheries for fish, crustaceans and molluscs become more fully exploited, so the effects of habitat destruction and pollution—particularly on those species depending on coastal wetlands and shallows or on inland wetlands and floodplains for nutrients or for spawning grounds and nurseries—will become more evident.

Preservation of Genetic Diversity

The preservation of genetic diversity is both a matter of insurance and investment—necessary to sustain and improve agricultural, forestry and fisheries production, to keep open future options, as a buffer against harmful environmental change, and as the raw material for much scientific and industrial innovation—and a matter of moral principle.

The issue of moral principle relates particularly to species extinction, and may be stated as follows. Human beings have become a major evolutionary force. While lacking the knowledge to control the biosphere, we have the power to change it radically. We are morally obliged—to our descendants and to other creatures—to act prudently. Since our capacity to alter the course of evolution does not make us any the less subject to it, wisdom also dictates that we be prudent. We cannot predict what species may become useful to us. Indeed we may learn that many species that seem dispensable are capable of providing important products, such as pharmaceuticals, or are vital parts of life-support systems on which we depend. For reasons of ethics and self-interest, therefore, we should not knowingly cause the extinction of a species.

Disappearing Cultivars

The genetic material contained in the domesticated varieties of crop plants, trees, livestock, aquatic animals and microorganisms—as well as in their wild relatives—is essential for the breeding programmes in which continued improvements in yields, nutritional quality, flavour, durability, pest and disease resistance, responsiveness to different soils and climates, and other qualities are achieved. These qualities are rarely if ever permanent. For example, the average lifetime of wheat and other cereal varieties in Europe and North America is only 5-15 years. This is because pests and diseases evolve new strains and overcome resistance; climates alter; soils vary; consumer demands change. Farmers and other crop-producers, therefore, cannot do without the reservoir of still-evolving possibilities available in the range of varieties of crops, domesticated animals, and their wild relatives. The continued existence of wild and primitive varieties of the world's crop plants is humanity's chief insurance against their destruction by the equivalents for those crops of chestnut blight and Dutch elm disease. This is not a remote.eventuality. It happened once with the European grape vine. In the 1860s *Phylloxera,* an insect which lives on the roots of the vine, arrived in Europe from North America. Its effect was catastrophic. Almost every vineyard on the continent was destroyed. Then it was discovered that the native American vine is tolerant of *Phylloxera.* Europe's wine production was saved only by the grafting of European vines onto American rootstocks—a practice that continues today.

The prospects of similar disasters striking other crops increase as farmers rely on fewer varieties. Because of intensive selection for high performance and uniformity the genetic base of much modern food production has grown dangerously narrow. Only four varieties of wheat produce 75% of the crop grown on the Canadian prairies; and more than half the prairie wheatlands are devoted to a single variety (Neepa-wa). Similarly, 72% of US potato production depends on only four varieties, and just two varieties supply US pea production.

Almost every coffee tree in Brazil descends from a single plant, and the entire US soybean industry is derived from a mere six plants from one place in Asia. These and other crops in a similar position are extremely vulnerable to outbreaks of pests and diseases and to sudden unfavourable changes in growing conditions. Unfortunately, while the genetic base of the world's crops and other living resources is narrowing rapidly, the means by which this dangerous situation could be corrected (the diversity of crop varieties and relatives) are being destroyed. Many wild and domesticated varieties of crop plants—such as wheat, rice, millet, beans, yams, tomatoes, potatoes, bananas, limes and oranges—are already extinct and many more are in danger of following them.

Valuable but primitive or locally distributed varieties are to an extent victims of their own utility, since the qualities of higher productivity and greater disease resistance that give the advanced varieties such an advantage over them are in large measure derived from them.

The rapid replacement of traditional varieties by new ones is a necessary and positive development given the need for more food; but it could prove counterproductive if the traditional varieties and their wild relatives are not saved as well. Primitive populations of crops and their wild relatives are an important source, and often the only source, of pest and disease resistance worth many millions of dollars, of adaptations to difficult environments and of other agro-nomically valuable characteristics such as the dwarf habit in rice and wheat, which has revolutionized their cultivation and led to greatly increased yields in many parts of the world.

Useful breeds of livestock are also at risk. Of the 145 indigenous cattle breeds in Europe and the Mediterranean region, 115 are threatened with extinction. Yet, as with crops, many traditional strains are of great value for breeding purposes. The very rare Wens-leydale sheep has been used to produce a heat tolerant breed able to produce good quality wool in subtropical lands; and the Cornish hen, once of interest only to poultry fanciers, proved so useful for crossing with other strains to produce a quick-growing meat bird that it is effectively the basis of the broiler industry.

Resources for Health

Although only a minute proportion of the world's plants and animals have been investigated for their value as medicines and other pharmaceutical products, modern medicine depends heavily on them. According to one analysis, more than 40% of the prescriptions each year in the USA contain a drug of natural origin—either from higher plants (25%) or microbes (13%) or from animals (3%)—as sole active ingredient or as one of the main ones. In the USA alone the value of medicines just from higher plants is reported to be about $3,000 million a year and rising. The most important applications of higher plants and animals for medicine are:

a. *as constituents used directly as therapeutic agents*—for example, digitoxin, morphine, and atropine, which are still unsurpassed in their respective fields;

b. *as starting materials for drug synthesis*—for example, adrenal cortex and other steroid hormones, which are normally synthesized from plant steroidal sapogenins;

c. *as models for drug synthesis*—for example, cocaine, which led to the development of modern local anesthetics. This application should not be underestimated. As one authority comments: "without naturally occurring active principles, it seems probable that neither the principle nor the activity would otherwise have been discovered. Put yourself in the place of a chemist who would like to develop a remedy for cardiac insufficiency; methods currently available would not lead him to synthesize a digitoxin-like molecule without knowledge of the natural prototype".

Paradoxically, a country's dependence on its own diminishing store of genetic diversity—as well as on that of other countries—is likely to grow as the country develops. Several developing countries, for example, are currently setting up their own pharmaceutical industries in order to supply their peoples with essential drugs at an acceptable cost. As a service to this effort a UN workshop recently compiled a basic list of medicinal plants found in Africa, Asia and Latin America whose active principles are used in modern medicine. More than 40 of the 90 species listed are available only from the wild; and another 20, though cultivated, are also taken from the wild. Preservation of these species and their habitats is thus one of the preconditions for maintaining indigenous pharmaceutical industries.

The history of human use of plant and animal species demonstrates that vanishing and apparently insignificant species can suddenly become useful, even important. The "pescado bianco" *Chirostoma estor,* a fish which in the wild occurs in a single Mexican lake, was until recently in danger of extinction as a result of overfishing, habitat degradation, and predation and competition by introduced species. Now, as a result of good management and artificial propagation, the fish is being stocked in several reservoirs and dams and a 15 hectare farm is under construction. Many species—for example the armadil lo and the polar bear—unexpectedly have been found useful for scientific research, whether as experimental material or as the providers of dues to technical innovations. Armadillos are the only animals other than human beings known to contract leprosy. They are now proving to be invaluable aids in the search for a cure for this disease. The recent discovery that polar bear hairs are exceptionally efficient heat absorbers has provided researchers with a clue that may help them design and produce materials for the manufacture of better cold-weather clothing and solar energy collectors.

Preservation of genetic diversity is thus necessary both to secure supplies of food, fibre, and certain drugs, and to advance scientific and industrial innovation. It is also necessary to ensure that the loss of species does not impair the effective functioning of ecological processes. It is unlikely that the particular communities of plants, animals and microorganisms that make up the ecosystems associated with so many essential processes—particularly pollination and naturally occurring pest control—can readily be substituted by other communities. The genetic composition of such ecosystems may be crucial for their performance.

Threats to Wild Species

Just as many varieties of domesticated plants and animals are disappearing, so too are many species of wild plants and animals. An estimated 25,000 plant species and more than a thousand vertebrate species and subspecies are threatened with extinction. These figures do not take account of the inevitable losses of small animal species—particularly invertebrates like molluscs, insects and corals—whose habitats are being destroyed. Indeed estimates that do attempt to take this factor into account suggest that from half a million to a million species will have been made extinct by the end of this century. If these species disappeared, the loss to humanity could be irreparable. The most serious threat is habitat destruction, which includes: replacement of the entire habitat by settlements, harbours and other human constructions, by cropland, grazing land and plantations, and by mines and quarries; the effects of dams (blocking of spawning migrations, drowning of habitat, alteration of chemical or thermal conditions); drainage, channelization and flood control; chemical nutrient and solid waste pollution (domestic, agricultural, industrial, mining); overex-traction of water (for domestic, agricultural and industrial purposes); removal of materials (such as vegetation, gravel and stones) for timber, fuel, construction and so on; dredging and dumping; overgrazing; and erosion and siltation.

The two other most serious threats to species are overexploitation and the effects of introduced exotic species. Exotic species, which may be introduced deliberately or inadvertently, can have adverse effects on native species in one or more of the following ways: competition for space or food; predation; habitat destruction or degradation; and transmission of diseases and parasites. The native species of fresh waters and of islands are particularly vulnerable to the harmful effects of introduced species. For example, introduced trout and bass are threatening many species of fishes in the USA; and introduced goats and rabbits are destroying the habitats of plants, birds and reptiles in the islands of the Pacific and Indian Ocean.

Sustainable Utilization of Species and Ecosystems

The necessity of ensuring that utilization of an ecosystem or species is sustainable varies with a society's dependence on the resource in question. For a subsistence society, sustainable utilization of most, if not all, its living resources is essential. So it is for a society (whether developing or developed) with a "one crop" or "few crop" economy, depending largely on a particular living resource (for example, the fishing communities of eastern Canada). The greater the diversity and flexibility of the economy, the less the need to utilize certain resources sustainably—but by the same token the less the excuse not to. Sustainable utilization is also necessary for the rational planning and management of industries dependent on the resources concerned (for example, timber, fish). Sustainable utilization is some what analogous to spending the interest while keeping the capital. A society that insists that all utilization of living resources be sustainable ensures that it will benefit from those resources virtually indefinitely. Unfortunately, most utilization of aquatic animals, of the wild plants and animals of the land, of forests and of grazing lands is not sustainable.

Aquatic Animals

On average, fish and other aquatic animals account for 6% of the total protein and 17% of the animal protein in the human diet >. If this seems small, it should be remembered that on a world basis most (65%) protein comes from plants—chiefly cereals, beans and peas, nuts and oilseeds. Meat accounts for 16% and milk products for 9.5% of the average total protein intake. These averages conceal substantial differences between and within countries. Aquatic animals are also important for trade. There are no world figures for domestic trade, but it is clear from export values alone that trade in seafood is both substantial and growing fast. In 1978 exports of fish and fishery products reached $10.8 thousand million, an increase of 15% over the previous year.

Because much utilization of fisheries is not sustainable, their contribution to national diets and incomes is likely to diminish. The result of past and present overfishing is that the annual world marine catch is 15-20 million tonnes (or about 20%-24%) *lower* than it might otherwise have been, and at least 25 of the world's most valuable fisheries are seriously depleted. The consequences of such overexploitation can be illustrated from the northwest Atlantic, where due to overfishing in the late 1960s cod catches are still only a third of their estimated potential. The drops in cod, herring and haddock catches (all caused by overfishing) could not be compensated by increased catches of capelin and mackerel; and the total catch of the fishery as a whole has declined from 4.3 million tonnes in 1970 to 3.5 million tonnes in 1976. It can no longer be assumed that depleted stocks will recover to their full potential, because: the spawning fishes and juveniles may continue to be caught by industrial fisheries (which take fish for conversion to animal feed); ecosystem dynamics can change and another species may take over because the depleted species can no longer compete effectively with it; and habitats essential for spawning or as nurseries may be degraded or destroyed.

Overfishing is the main threat to marine living resources and a significant threat to freshwater ones. It occurs locally in all regions, but it is generally most pronounced in regions dominated by developed countries.

Five of the eight regions with most stocks that are depleted are developed (northwest Atlantic, northeast Atlantic, Mediterranean, northwest Pacific, northeast Pacific). Of the remaining three, two (eastern central Atlantic and southeast Atlantic) are dominated by developed country fishing fleets: France, Japan, Poland, Spain, USSR, South Africa— with Cuba and the Republic of Korea operating the only major developing country fleets. The only developing region dominated by developing country fleets is the southeast Pacific, where most of the fishing is done by Peru and Chile 1

As well as depleting many fish and a few mollusc stocks, overfishing has almost extinguished entire species of whales, sea cows and sea turtles. Many aquatic animal groups are also under pressure because of accidental overexploitation or "incidental take". The incidental capture and killing of non-target animals in the course of hunting or fishing for other species is one of the more destructive yet neglected problems of aquatic living resource management. It is not only highly wasteful, destroying an estimated 7 million tonnes of fish every year, it also threatens the survival of several sea turtle species, notably the Kemp's ridley turtle which has been reduced almost to extinction. A million seabirds are killed accidentally in nets every year; and more cetaceans, notably dolphins and porpoises, are taken incidentally than deliberately.

Wild Plants and Animals of The Land

Wildlife is an important subsistence resource in developing countries and an important recreational resource in both developed and developing countries. Many wild plants and animals of the land are an important renewable resource and source of food, particularly for rural communities in developing countries. In parts of Ghana, Zaire and other countries in west and central Africa, for example, up to three-quarters of the animal protein comes from wild animals,. The nutritional importance of wild animals and plants for large numbers of people is invariably underestimated (and often ignored), largely because many of the more frequently eaten plants and animals (for example, vetches and porcupines) seldom feature in the diets of nutritionists and are harvested in areas far from the scrutiny of statisticians. This is unfortunate, for were the true nutritional value and the use made of wild plants and animals appreciated by governments they might be more ready to encourage these resources to be managed sustainably and to take steps to retain their habitats.

Wild animals and plants also provide a significant, and sometimes the only, source of income for rural communities. In Canada, for example, the wild fur trade helps support 40,000 trappers, who in the 1975-1976 season. caught $25 million worth of pelts, mostly from beaver, muskrat, lynx, seal, mink and fox. By comparison, in the same season, farms produced $17 million worth of furs, 99% being mink. World trade in wildlife and wildlife products has become big business: in 1975 the USA alone imported more than $1,000 million of wildlife products.

International trade has become a threat to many species as well-organized commercial enterprises attempt to supply a vastly expanded market (mainly industrialized countries) with increasingly scarce "commodities" taken from the wild (mainly in developing countries):

— hides and skins for the luxury fur and leather industry; exotic meat and fish for luxury food; a wide range of other animal and plant products for pharmaceuticals, perfumes, cosmetics, aphrodisiacs, decoration, souvenirs or investment; specimens for natural history museums;

— live plants for horticulture; live animals for the pet trade, zoos and menageries, aquaria and other collections; and for the testing of new chemical products and for biomedical research.

— Most of the trade is openly advertised, but a significant part of it takes place illegally, often through channels and by methods not unlike those of the drug traffic. The impact of this trade on many species and ecosystems is now serious. For example, overexploitation threatens almost 40% of all vertebrate species in danger of extinction, and is the most serious of the threats faced by reptiles. Because of the difficulty of securing supplies, world trade in crocodilian hides has dropped from an estimated 10 million hides to 2 million.

Finally, wildlife is a major resource base for recreation and tourism. Tourism, largely based on wildlife, is among Kenya's top three foreign exchange earners. In Canada 11 % of the population hold hunting licences; in the USA 8% hold hunting licences and 13% hold fishing licences; and in Sweden from 12% to 18% hold fishing licences. Many more people enjoy simply looking at wildlife: in the USA there are about 7 million birdwatchers, 4.5 million wildlife photographers, and almost 27 million nature hikers. For a great many people, too, wildlife is of great symbolic, ritual and cultural importance, enriching their lives emotionally and spiritually.

Forests and Woodlands

Forests and woodlands provide a rich variety of goods, useful to affluent industrial and poor rural communities alike: timber, sawnwood and panels for construction, walls, doors, shuttering and furniture; pulpwood for pulp, paper, cartons and rayon; poles, posts, mining timbers and railway track sleepers; fuelwood; fodder, fruits, game meat, honey, pharmaceuticals, fibres, resins, gums, dyes, skins, waxes and oils; beauty, amenity and recreation. Forests have unquestioned importance for industry and commerce. The value of the annual world production of forest products exceeds $115,500 million, and international trade is worth about $40,000 million a year. Thirty countries (eight of them developing countries) each earn more than $100 million a year from exports of forest products—and five of these each earn more than $1,000 million a year.

More than 1,500 million people in developing countries depend on wood for cooking and keeping warm. Their annual consumption of wood is estimated to be more than 1,000 million m, well over 80% of developing countries' total wood use (excluding exports). In Africa the contribution of trees to total energy use is as high as 58%; in Southeast Asia and Latin America it is 42% and 20% respectively. The effect of such intense demand is to denude the land of wood over wide areas. Around one fishing centre in the Sahel region of Africa, where the drying of 40,000 tonnes offish consumes 130,000 tonnes of wood every year, deforestation extends as far away as 100 km. Fuelwood is now so scarce in the Gambia that gathering it takes 360 woman days a year per family. Even when firewood is available for sale, it is often beyond the budgets of poor householders. In the highlands of the Republic of Korea cooking and heating can cost up to 15% of the household budget; and in the poorer parts of the Andean Sierra and the Sahel it can be as high as 25%. Consequently many families are forced to do without.

Grazing Lands

Permanent pastures (land used for 5 years or more for herbaceous forage crops, whether cultivated or wild) are the most extensive land-use type in the world—occupying 30 million km, or 23% of the earth's land surface. Permanent pastures and other grazing land are generally in areas of low and irregular rainfall and are usually unsuitable for crops without intensive capital investment. Their productivity is generally low, ranging from 1 hectare supporting 3-5 animal units on fertile, well-managed pastures in central Europe to 50-60 hectares to support 1 animal unit in Saudi Arabia. Nonetheless, grazing lands and forage support most of the world's 3,000 million head of domesticated grazing animals, and hence most of the world's production of meat and milk'.

Unfortunately, mismanagement of grazing lands is widespread. Overstocking has severely degraded grazing lands in Africa's Sahelian and Sudanian zones. In parts of North Africa, the Mediterranean and the Near East, it is a major contributor to desertification. In many areas farmers are moving onto land that is marginal for agriculture, thereby displacing pastoralists onto land that is marginal for livestock rearing. Overstocking together with uncontrolled grazing is also a serious problem in mountain areas, such as the Himalaya and the Andes. There, too many improperly tended animals remove both trees and grass cover—which is often very poor, both as forage and for soil protection—and erosion accelerates.

Ecological Processes and Life-suuport Systems

Reserve Good Cropland Crops: In view of the scarcity of high quality arable land and the rising demand for food and other agricultural products, land that is most suitable for crops should be reserved for agriculture. This will reduce the pressure on ecologically fragile marginal lands which tend to degrade rapidly if exploited beyond their productive capacities. However, this requirement may conflict with urban, industrial, energy or transport policy. There are many examples of prime farmland drowned by dams or lost to airports, roads, factories or housing. Without careful planning and zoning, human settlements sited in farming areas are bound to encroach on farmland as they expand. Such conflict should be anticipated and where possible avoided. Since it is not possible to resite high quality cropland but it is possible to be flexible about the siting of buildings, roads and other structures, agriculture as a general rule should have precedence.

The requirement to reserve good cropland for crops may also conflict with other conservation needs. There are unmodified ecosystems with great agricultural potential which should be protected to preserve genetic diversity and as a control area for baseline monitoring and scientific research. Again, agriculture generally should have precedence, but *only* in the case of land with no serious limitation for agriculture, since such land is at a premium especially in developing countries. In countries where land "most suitable" for agriculture nonetheless has moderately severe limitations for agriculture, it is necessary to protect some areas as controls so that the long term effects of agricultural and other activities on such land can be as sessed. Floodplains and wetlands are special cases. When drained or protected from inundation they can provide productive agricultural land. However, such conversion may deprive important fisheries of essential support and may also lead to the loss of other valued living resources. Where floodplain agriculture is practised the losses of agricultural production and of fisheries may not be compensated by production from the irrigated agriculture that generally replaces them. The social, economic and ecological costs and benefits of convert ing floodplains and wetlands need to be thoroughly assessed, therefore, before conversion is permitted.

A General Note on Priority Requirements

Manage cropland to high, ecologically sound standards. This requires soil and water conservation, the recycling of nutrients, and retention of the habitats of organisms beneficial to agriculture. To avoid aquatic pollution and to conserve inorganic fertilizers, crop residues and livestock wastes should, as far as possible, be returned to the land. To maintain and promote the beneficial contribution of crop pollinators and of the natural enemies of pests to integrated pest control, the habitats necessary for the organisms concerned should as far as possible be retained. In intensive production systems where the habitats of the required organisms are too few to maintain them in adequate numbers, the organisms should be propagated artificially. Chemical fertilizers and pesticides will still be needed, but they should be used with care and should supplement rather than supplant good nutrient, soil and pest management practices.

Many tropical soils quickly lose their fertility. Traditional systems of shifting cultivation restored fertility by leaving the land fallow for long periods, but fertilizers are indispensable for continuous cropping. Manufactured fertilizers are beyond the means of many developing country farmers because of their high cost, low prices for farm products, shortage of credit and a lack of fertilizer supplies. Since chemical fertilizers are derived from petroleum their cost is likely to continue to rise sharply with the growing cost of oil. The estimated 113 million tonnes of plant nutrients that are potentially available to developing countries from human and livestock wastes and from crop residues should as far as possible be used to fertilize the land. Properly managed, organic wastes could help substantially to increase agricultural production, restore and maintain land in good condition, and reduce aquatic pollution. In many areas, even when inorganic fertilizers are readily available, they should be used only in combination with organic fertilizers. For example, the dominant kaoli-nitic clay soils of the humid tropics have low absorption capacity, and chemical fertilizers need to be combined with organic manures to be fully effective. The use of organic wastes as plant nutrients and soil restorers can be combined with the production of biogas (methane). This process eases the problems of storing and delivering organic wastes, reduces the loss of organic matter through decomposition, and provides gas for domestic uses.

If land is eroding so rapidly that it must be retired, there is seldom any alternative to the generally difficult tasks of resettling the farmers concerned else-where or of absorbing them into other sectors of the economy. Situations where land retirement is the only solution can be avoided by promoting systems of production adapted to ecological conditions in which modern technology and techniques are integrated with traditional systems of resource management. This is particularly important for communities whose shifting cultivation practices have become unstable because the rising population demands more intensive production than the soil can support without considerable improvement. The techniques and inputs of sustainable permanent cropping—such as chemical fertilizers, improved seed, and soil conservation measures—are usually beyond the economic means of poor farmers. But it is possible to shorten the fallow period of shifting cultivation gradually through mixed cropping practices, the limited use of inorganic fertilizers, and recycling organic materials. It is also possible to improve the efficiency of the fallow period by replacing natural cover with cover crops which can be used economically—such as pasture for livestock in mixed agri-pastoral systems, or tree crops in mixed agri-silvicultural systems.

Ensure that the principal management goal for watershed forests and pastures is protection of the watershed. This is particularly important in the upper catchment areas where rivers originate and where often rainfall is greatest. Especially fragile or critical areas, notably steep slopes with erodible soils and the source areas of major rivers should never be cleared. In other areas, uses compatible with watershed protection may be permitted provided the government is able to ensure that the primary management goal is respected. For example, protective forest may be converted to a tree crop (either for timber or for some other product, such as a tea plantation), if great attention is paid to hydrological control. This demands high capital input, professional skill and great conscientiousness. If the regulatory authority feels any of these requirements is lacking, the forest should not be altered. Where watershed forests have already been severely reduced, and siltation and flooding have increased, efforts should be made to restore the ecosystem by reforestation rather than to resort to channelization or impoundment. Sometimes, however, a government will have little room for manoeuvre—as when a watershed forest is already being converted to agriculture by local communities. Reforestation and other land rehabilitation measures, matched by conservation-based rural development and intensification of production from better land, will then be needed.

Ensure that the principal management goal for estuaries, mangrove swamps and other coastal wetlands and shallows critical for fisheries is the maintenance of the processes on which the fisheries depend. Other uses of these ecosystems should not impair their capacity to provide food and critical habitat for economically and culturally important marine species. Similarly, major management goals for coral ecosystems and active floodplains (where they are important for fisheries) should be maintenance of the fisheries concerned. Fishing at sustainable levels using nondestructive methods (not, for example, dynamite) is of course compatible with this goal, and so too are recreation and tourism at sustainable levels. An additional major goal of coral reef management will often be the preservation of genetic diversity. The management of active floodplains is likely to be complicated by many competing uses which may be as or more important than fisheries : for example, agriculture and livestock raising. However such uses need not be incompatible.

Control the discharge of pollutants. The discharge of pollutants and use of pesticides and other toxic substances should be controlled. Care should be taken to avoid contamination of the habitats of threatened, unique or economically important species. Special at tention should be paid to substances that are highly toxic, or are released in large quantities, or persist in the environment and accumulate in living organisms. The impacts of such substances on ecosystems and species should be monitored, regularly evaluated, and reduced to levels that can be tolerated by the ecosystems and species concerned. Since the effects on ecosystems and species of the thousands of chemicals that enter the environment are largely unknown, general surveillance of the environment should be undertaken.

Genetic Diversity

Prevent the extinction of species

Priority should be given to species that are endangered throughout their range and to species that are the sole representatives of their family or genus, according to the following formulation: the greater the potential genetic loss the less imminent that loss need be to justify preventive action.

Families and genera that are mono-typic (consist of only one species) should receive priority over polytypic ones, since—theoretically—the smaller the family (or genus) the greater the gap between the nearest related family (or genus) and therefore the more distinct that group of species is from others. Other things being equal, an endangered species should be given priority over a vulnerable one; a vulnerable over a rare one; and a rare species over one that even if it is declining is considered insufficiently threatened to qualify for one of the three categories. Imminence of threat, however, is partly a matter of the state of knowledge of a species. Species not known to be threatened but with highly restricted distributions should therefore be closely monitored—with particular attention being paid to higher taxa (families and genera).

Prevention of extinction requires sound planning, allocation and management of land and water uses, supported by on site *(in situ)* preservation in protected areas and off site *(ex situ)* protection such as in zoos and botanical gardens. Protected areas can preserve more wild species, subspecies and varieties than can off site protection; but to be fully effective both must be integral components of rational resource man- agement. Such management should include protection from threats other than habitat destruction or degradation, notably overexploita-tion (both deliberate and incidental) and the effects of introduced exotic species. These measures may be assisted by participation in international programmes for the prevention of species extinction. Where an introduced exotic species is having adverse effects on native species, the introduced species should be eliminated if possible. Given the extreme difficulty of eliminating introduced species, however, every effort should be made to prevent all introductions except those which, before the introduction is made, can be shown to provide economic, social and ecological benefits substantially greater than any costs, and over which adequate control can be exercised. A proposed introduction should be the subject of an environmental assessment, including a full enquiry into the likely and possible ecological effects.

Preserve as many varieties as possible of crop plants, forage plants, timber trees, livestock, animals for aquaculture, microbes and other domesticated organisms and their wild relatives. Priority should be given to those varieties that are most threatened and are most needed for national and international breeding programmes. This requires both off site and on site preservation and may be assisted by participation in international programmes for the preservation of genetic resources.

There are three ways of preserving genetic diversity:

a. *on site*—in which the stock is pre-served by protecting the ecosystem in which it occurs naturally;

b. *off site, part of the organism*—in which the seed, semen or other element from which the organism concerned can be reproduced is preserved;

c. *off site, whole organism*—in which a stock of individuals of the organism concerned is kept outside its natural habitat in a plantation, botanical garden, zoo, aquarium, ranch or culture collection. Of the three, the preferred measure is on site preservation. However, this is not possible for domesticated species; and even wild species should be preserved off site as a safeguard against any failure in on site preservation.

Off site preservation involves the following steps: exploration of the remaining diversity of the target species; collection, giving priority to materials likely to be lost if nothing were done; preservation in storage (part of the organism) or maintenance in a plantation, culture collection, and so on (whole organism); documentation, requiring the systematic description of collected materials (especially their place of origin and their taxonomic and morphological features) and the recording, organization and retrieval of this information; evaluation of agronomic qualities (such as yield potential, cooking and nutritional qualities), biotic qualities (such as pest and disease resistance), and eco-edaphic qualities (such as drought resistance, temperature tolerance, responsiveness to different soil conditions); and utilization, combining genes from various sources into improved strains or varieties.

Special attention should be paid to the preservation of genetic material for forestry and for aquaculture. If forestry and fisheries are to make as complete a transition to domestication as crop production has done, the genetic base of the two industries must be preserved. This requires accelerated programmes of off site preservation in protected areas, coupled with inventories of existing protected areas, documenting, evaluating and utilizing the materials within them. Microorganisms are a special case, because of their vast numbers, great resilience and ability to adapt to environmental change, and their very rapid rates of reproduction. Preservation of microbial strains is necessary not so much to prevent extinction but to aid utilization. Isolating from nature a strain with a particular desired property is tedious. Once isolated, therefore, such strains need to be maintained in culture collections.

Ensure that on site preservation programmes protect: the wild relatives of economically valuable and other useful plants and animals and their habitats; the habitats of threatened and unique species; unique ecosystems; and representative samples of ecosystem types. Inventories of existing protected areas should be made to determine what threatened, unique and other important species may already be protected adequately. Each country should identify the habitats of such species, and ensure their preservation in protected areas as a matter of priority. Whenever feasible, each protected area should safeguard all the critical habitats (the feeding, breeding, nursery and resting areas) of the species concerned. Where this is clearly not feasible—as in the case of migratory or wide-ranging animals—a network of protected areas should be established, the effect of which would be to safeguard all the habitats of the species concerned. If the species migrates or ranges from one national jurisdiction to others, bilateral and multilateral agreements should be made as appropriate to set up the required network. Other uses of protected areas may be permitted provided they are compatible with protection of the habitats concerned. Not only should habitat be protected but any external source of the nutrients and other essentials on which each habitat depends should also be protected or so managed as to assure an adequate supply of the essential concerned. Exploitation and other impacts (such as pollution) along migration routes should also be regulated.

Unique ecosystems should be protected as a matter of priority. Only those uses compatible with their preservation should be permitted. In addition, a complete range of ecosystems representative of the different types of ecosystem in each country should be protected so that the range of variation in nature is preserved. Only those uses that are compatible with the preservation of the ecosystem and its component communities of plants and animals should be permitted in areas protected for these purposes. Each country should review its existing system of protected areas and ascertain the extent to which the different kinds of ecosystem in each biogeographical province are protected. Biogeographical provinces with no protected areas should be given priority, followed by the provinces in which few of the ecosystem types are represented in protected areas. Attention should be paid to the adequacy of protection of each area. Global biogeographical classifications should be used together with more detailed national or regional classifications derived from them. Special attention should be paid to marine ecosystems which are particularly poorly represented in protected areas. Priority should be given to ecosystems that are particularly rich in species or have no adequately protected areas.

Determine the size, distribution and management of protected areas on the basis of the needs of the ecosystems and the plant and animal communities they are intended to protect. Generally, a large reserve is better than a small one.

Areas chosen for protection should have as much internal variation as possible. The necessary measures should be taken to safeguard the support systems of protected areas and to shelter the areas from harmful impacts: these measures should include the establishment of buffer zones where special restraints on use may be applied. Research should be continued into the questions of minimum critical size and optimum distribution of the protected areas required to safeguard a given number and composition of species. Security for protected areas must be clearly provided for in national legislation.

Coordinate national protected area programmes with international ones, particularly the biosphere reserves programme of Unesco's Man and the Biosphere Project 8 and the initiatives of IUCN's Commission on National Parks and Protected Areas, so that a complete network of protected representative samples of ecosystems may be established as soon as possible. One of the njajor objectives of the international network of biosphere reserves is to conserve for present and future use the diversity and integrity of plant and animal communities within natural ecosystems, and to safeguard the genetic diversity of species on which their continuing evolution depends.

Countries are urged to contribute to this network by designating sites representative of the biogeographical provinces or other major ecosystem groups found in their territory.

Sustainable Utilization

1. *Determine the productive capacities of exploited species and ecosystems and ensure that utilization does not exceed those capacities.* Species and ecosystems should not be so heavily exploited that they decline to levels or conditions from which they cannot easily recover. Such levels and conditions vary depending— in the case of a species— not only on the biology of the species but also on the quality of the ecosystems that support it. If the ecosystems are being altered by human activities (such as exploitation of associated species, pollution, incidental take) then the level from which the population is unable to recover could be substantially higher than the level presumed from study of the species alone.
2. *Adopt conservative management objectives for the utilization of species and ecosystems.* Management objectives should take adequate account of important relationships between the exploited species or ecosystems and the species and ecosystems with which they are linked. They should allow for error, ignorance and uncertainty. When a single species (rather than a group of species or an ecosystem) is exploited, and the species is at the top of the food chain, stocks should be kept at or above the level at which they provide the greatest net annual increment. If the species is not at the top of the food chain, stocks should not be depleted to a level such that the population's productivity, or that of the populations of other species dependent on it, is significantly reduced. When a group of species is exploited, catch levels should be fixed so that the productivity of the species with the slowest recovery time (often predators at the top of the food chain) is not significantly reduced.
3. *Ensure that access to a resource does not exceed the resource's capacity to sustain exploitation.* Measures to regulate utilization can include: restricting the total take, the number of persons, vessels or other units allowed to participate in exploitation and the times and places of exploitation; and prohibiting or restricting the use of certain methods and equipment. Of these, a combination of quotas and restrictions on access to the resource is usually essential. Quotas alone

are highly vulnerable to excessive increase in response to political pressure by vested interests, if those interests are allowed to grow to a size which the species or ecosystem cannot support.

4. *Reduce excessive yields to sustainable levels.* Industries, communities and countries that are overexploiting living resources on which they depend would be better off in the medium and long term if they voluntarily reduced yields to levels that are sustainable. In this way they could adjust to realistic levels of consumption and trade and avoid unexpected drops in yield, instead of being surprised by them when they occurred. The economic distress that involuntary, unscheduled cuts may cause is avoidable by the planned reduction of utilization to sustainable levels.

5. *Reduce incidental take as much as possible.* This can be done through the establishment of protected areas or of closed seasons (prohibiting fishing in at least one self-sustaining area or at times when the affected species is particularly vulnerable, in the case of sea turtles, for example, during both the nesting and the hibernating periods); or through modifications of fishing gear or methods. Many of the commercially exploited marine and freshwater fish communities, however, occur in species complexes. It is impossible to exploit such complexes intensively without disturbing the species ratios and threatening at least some species with severe depletion. In such cases, reserves should be established in which commercial exploitation is prohibited. Subsistence exploitation may still be compatible with species protection, depending on fishing intensity and methods.

6. *Equip subsistence communities to utilize resources sustainably.* Where a community depending for subsistence wholly or partly on living resources effectively regulates utilization so that it is sustainable, its regulatory measures should be supported. Where there is no regulation or where traditional regulatory measures have been rendered obsolete (by, for example, growth in the community's population, the advent of more destructive methods of exploitation or of commercial exploitation), the community should be helped to devise and enforce a set of effective regulations. Should overexploitation have become so acute that very severe regulation is needed (including an outright ban of months or years on exploitation) efforts should be made to gain the understanding of the community and its participation in enforcement;

and an alternative food, fuel or fibre should be offered. Commercial utilization and trade should be prohibited before the subsistence take is reduced.

7. *Maintain the habitats of resource species.* Where agriculture can supply more food, more economically and on a sustainable basis, than can the utilization of wildlife, the conversion of wildlife habitat to farmland is rational. Often, however, the habitats of wildlife are destroyed for short-lived agricultural and other developments with a net loss in welfare for the local communities. Especially in areas with severe limitations for agriculture, the value of wildlife should be carefully assessed and the returns on habitat management for more intensive, yet sustainable, wildlife utilization compared with the returns on the destruction of habitat to enable some other use. This requirement is most important in tropical forest areas, where rural communities often depend on wildlife for a high proportion of their protein as well as for other goods. It is also important to maintain the freshwater, coastal and marine systems that supply food and shelter to aquatic animals, especially those that support fisheries. This requires control of human impacts in river basins (including the upper catchment area) as well as in the coastal zone and at sea.

8. *Regulate international trade in wild plants and animals.* The most promising way of doing this is through the Convention on International Trade in Endangered Species of Wild Fauna and Flora (CITES). CITES now has a worldwide membership of more than 50 states. To make its system of export/import controls fully effective other trading nations should join. Those that have become members should take all necessary administrative measures to enforce the Convention at the national level. Species threatened by overexploitation that are not yet covered by CITES should be, if they are likely to enter international trade.

9. *Allocate timber concessions with care and manage them to high standards.* The siting and management of timber operations should be such that essential processes (especially watershed protection) are maintained. Unnecessary damage to trees that are not utilized should be avoided. Felling programmes should be matched by planting programmes, as far as possible using the species exploited, so that what is taken out is replaced. Most timber companies are capable of taking all the necessary measures and should undertake to do so. It is advisable

that governments equip themselves to inspect and control the conduct of logging operations before they take place. Similarly, if there is any likelihood that spontaneous settlement will occur in a forest area once it has been opened up for logging, governments should ensure they have a practicable plan for assisting settlers to develop the land sustainably, for providing firewood plantations or alternative fuels, for securing essential processes and for protecting important genetic resources.

10. *Limit firewood consumption to sustainable levels.* Where vegetation is being destroyed by cutting and stripping for fuel immediate measures should be taken to:
    — establish plantations for firewood, large enough to meet higher levels of demand than today's;
    — provide alternative sources of firewood, to take pressure off the plantations and remaining vegetation;
    — restore the vegetation;
    — provide stoves that utilize firewood more efficiently;
    — provide alternative sources of energy (such as biogas).

11. *Regulate the stocking of grazing* lands so that the long term productivity *of plants and animals can be maintained.*

    The carrying capacity of grazing lands is determined by the annual production of plant growth in excess of what is required by the plants for their metabolism, health and vigour. Much of this excess production can be cropped by wild animals or livestock without damage to the vegetation. Careless or excessive use, however, impairs the plants' capacity to grow and reproduce. This in turn leads to sometimes permanent destruction of the vegetation or to a decline in the proportion of plants palatable to livestock or both. In arid regions, where rainfall and plant growth are erratic, stocking densities must be more conservative than where annual productivity is more consistent. In such regions nomadism and transhumance (the seasonal movement of livestock) are often the most sustainable strategies and, if still practised, should not be abandoned without good reason.

12. *Utilize indigenous wild herbivores, alone or in combination with livestock, where the use of domestic stock alone will degrade the land.* Native wild herbivores are adapted to make use of natural grazing land without deterioration. In extreme conditions they may be the only

species that can do so; and elsewhere they can provide an economically, ecologically and socially desirable alternative or supplement to domestic livestock. The potential of wild herbivores for subsistence and commercial use should be given priority attention. Two main actions are needed:

— assessment of social and economic potential of game ranching, looking at commercial utilization, subsistence utilization, and domestication options, as well as at market potential for products;
— assessment of current and potential ecological impacts of trypanosomiasis control in Africa, including consideration of new developments in control techniques.

Priorities for National Action

A Framework for National and Subnationational Conservation Strategies

1. To ensure that the objectives of conservation are achieved as expeditiously as possible and to speed the integration of conservation with development, it is recommended that every country review the extent to which it is achieving conservation, concentrating on the priority requirements and on the main obstacles to them. The review should form the basis of a strategy to overcome the obstacles and meet the requirements. The strategy may be at the national level or at one or more subnational levels (provincial, state, municipal), or there may be separate (but, it is hoped, complementary) strategies at several levels, depending on the division of government responsibilities for the planning and management of land and water uses.
2. The purpose of these strategies is to focus attention on the relevant priority requirements for conservation, to stimulate appropriate action, to raise public consciousness, and to overcome any apathy or resistance there might be to taking the action needed. National and sub-national strategies are intended to provide a means of focussing and coordinating the efforts of government agencies, together with nongovernmental conservation organizations, to implement the World Conservation Strategy within countries.
3. Although the planning and execution of conservation strategies is primarily the responsibility of governments, nongovernmental

organizations should be fully involved to ensure that all the resources available to conservation are deployed coherently and to the full and thus to accelerate the achievement of. conservation objectives. Indeed in some countries nongovernmental organizations may wish to take the initiative.

Steps in a Strategy

4. It is recommended that each strategy proceed as follows:

   a. review development objectives in relation to each of the conservation objectives; and describe the extent to which each of the conservation objectives is or is not being achieved— with particular emphasis on the priority requirements—and the status of and threats to the living resources concerned;

   b. identify the main obstacles to achieving the objectives and to removing or reducing the threats; and identify any special opportunities' there may be for overcoming such obstacles;

   c. identify the measures required to achieve the objectives and to remove or reduce the threats to the living resources concerned;

   d. determine priority ecosystems and species, the requirements for their conservation, and how these requirements could be met—providing planning and management guidelines with respect to such ecosystems and species;

   e. analyze present and planned activities, comparing them with c and d, and identify gaps that need filling and activities that need strengthening and supporting;

   f. estimate the, financial and other resources, and the legislative and administrative measures, required to carry out the actions identified in e, and identify the organizations that should be carrying them out;

   g. propose ways of supplying the financial and other resources required and of authorizing and equipping the appropriate organizations to carry out the required actions, identifying the bodies that possess the necessary resources and powers of decision;

   h. set out a plan of action to bring about the required political decisions and allocations of financial and other resources;

i. set out a programme of required measures, including administrative and legislative measures, for the maintenance of essential ecological processes and life-support systems, the preservation of genetic diversity, and the sustainable utilization of ecosystems and species, noting particularly those priority requirements not yet met.

5. Irrespective of its purpose, every strategy has certain functions: to determine the priority requirements for achieving the objectives; to identify the obstacles to meeting the requirements; to propose the most cost-effective ways of overcoming those obstacles.

   When resources are limited and time is running out, it is essential to be sure that the available resources and effort are applied to the highest priority requirements first, and only afterwards to lesser priorities. Conservation is in exactly this situation, yet conservation organizations have seldom attempted to agree priorities. This is an understandable failing, since there are so many urgent problems to be dealt with, people have different perceptions of priorities, and there have been few universally accepted criteria for what is important. However, it is precisely because there are so many requirements—most of them urgent, and many of them alone demanding.all or more of the resources at conservation's disposal—that priorities must be determined and followed.

Strategic Principles

6. In addition, strategies to meet the priority requirements for the achievement of conservation objectives should take into account these four strategic principles:

Integrate. The separation of conservation from development together with narrow sectoral approaches to living resource management are at the root of current living resource problems. Many of the priority requirements demand a cross-sectoral, interdisciplinary approach.

Retain options. Our understanding of the dynamics and capacities of many ecosystems, particularly tropical ones, is often insufficient to assure rational use allocation or high quality management. Scientific knowledge of the productive capacities of most tropical ecosystems, as well as of their ability to absorb pollution and other impacts, is generally inadequate. Land and water use, therefore, should be located and managed so that as many options as possible are retained.

Mix cure and prevention. Current problems are often so severe that it is tempting to concentrate on them alone; impending problems could be still worse, however, unless early action is taken to prevent them. Strategies for action should therefore be a judicious combination of cure and prevention—of tackling current problems and of equipping peoples and governments to anticipate and avoid future problems.

Focus on causes as well as symptoms. When conservation puts itself into the position of dealing only with symptoms it appears unduly negative and obstructive. A late attempt to modify a development, whether successful or not, comes across as anti-development (hence anti-people) even though this is seldom the case. The result is either an outright defeat or, because it generates hostility and mis-. conceptions, a victory that has within it the seeds of future defeats. Furthermore, by the time symptoms appear it is often too late to do anything about them, because many ecologically unsound projects are the results of already fixed policies and part of complex and expensive plans the proponents of which are understandably reluctant to unravel. This said, it is also important not to neglect the symptoms. Although interventions are more effective the earlier in the development process they are made, in practice they are needed at all stages. In addition it is sometimes not possible to deal with causes, since many of them are complex and beyond the capabilities of conservation organizations to influence. Action directed at causes generally yields results only over the long term. Symptoms may be so acute that action must be taken immediately.

Main Obstacles

7. The most effective way of preventing irreversible damage to living resources is through overcoming the main obstacles to achieving the requirements of conservation.:
    a. absence of conservation at the policy making level;
    b. lack of environmental planning and of rational use allocation;
    c. poor legislation and organization
    d. lack of training and of basic information;
    e. lack of support for conservation;
    f. lack of conservation-based rural development;

8. The importance of these obstacles cannot be overemphasized. Few if any countries take adequate account of ecological considerations when making policy or planning development. Few allocate or regulate uses of their living resources so as to ensure that they are ecologically appropriate and sustainable. Many lack either the financial or technical resources, or the political will, or adequate legislative, institutional or public support for conservation (or any combination of these) to carry out fully the conservation measures required. The result is that the number of urgent conservation problems proliferates. A species may be rescued, an area protected, or an environmental impact reduced, but such successes will be temporary or will be overshadowed by much greater failures unless every country's capacity to conserve is greatly improved and permanently strengthened.
9. Special attention needs to be paid to the ability of government agencies to deal not only with current problems but also with potential ones since they may often be beyond the capacities, man dates and experience of the bodies responsible. The mandates, capacities and procedures of the government agencies and other bodies responsible for development and conservation should be analyzed to assess the extent to which ecological considerations are incorporated into the development process and whether conservation laws and organizations are strong enough to ensure that the required conservation measures are carried out. The analysis should focus on the factors which help or hinder meeting the priority requirements for conservation. Then, particular strengths and weaknesses should be brought to the attention of the public, legislators and the policy making level of government, together with specific proposals for filling any serious gaps in planning, decision making and management.
10. Although most conservation progress is to be won within government bodies and other organizations concerned with development and conservation, the measure of such progress will be improvements on the ground and in the water: more secure and wide spread maintenance of essential ecological processes and life-support systems, preservation of genetic diversity, and sustainable utilization of species and ecosystems. The status of key species and ecosystems needs therefore to be closely monitored, and strategies adjusted in the light of any improvement, deterioration or absence of change. Strategies are means,

not ends in them selves. But the process by which they are forwarded is itself usually of value, as it can inform and educate, develop participation in and support for decision making, change attitudes, and help to foster a conservation ethic.

Policy Making and The Integration of Conservation Development

Development that is inflexible and little influenced by ecological considerations is unlikely to make the best use of available resources. By causing ecological damage it is likely also to cause economic and social damage. The most effective way society can avoid such problems is to integrate every stage of the conservation and development processes, from the initial setting of policies to their eventual implementation and operation.

The problems

To achieve the objectives of conservation governments need to dispel any notion that conservation is a limited, independent sector largely concerned with wildlife or with soil; and that ecological factors are impediments to development which in some cases may safely be overlooked and in others may be considered simply on a project by project basis, not as a matter of policy. Unfortunately these beliefs are implicit in the way policies generally are made and operated. Such a narrow interpretation of conservation has at least three important consequences. First, the ecological effects of a particular development policy are seldom anticipated and hence the policy is not adjusted in time to avoid expensive mistakes. Second, those sectors directly responsible for living resources (notably agriculture, forestry; fisheries, and wildlife) are often impelled to concentrate on production at the expense of maintenance, with the result that otherwise renewable resources are dissipated and the resource-base of future utilization is undermined. Third, because of a previous lack of conservation, the policies of other sectors may be frustrated. The energy sector's forecasts of the life of a hydroelectric power station, for example, may be completely falsified by poor watershed management.

Even when ecological factors are considered, it is seldom at the critical policy making stage when the basic pattern of development is often fixed. Consideration at the project stage, although often necessary, is no substitute for proper consideration at the policy stage—for by the project stage economic and social requirements will normally have been set so firmly that only minimal or cosmetic adjustments are possible. If, however, a bold decision is taken to uphold conservation and arrest an ecologically unsound scheme, it may be at the cost of major confrontations with vested interests, social conflict and a waste of human and financial resources.

The bias of living resource agencies towards production rather than maintenance is often a response to competition within governments for scarce financial resources, and the consequent pressure on all sectors to show results that can be directly related to economic performance. Under the circumstances, agencies with the dual task of regulating and promoting resource development are likely to find it difficult to balance the two requirements. This difficulty is exacerbated by the lack of a well-defined, generally agreed measure of conservation performance. Economic performance can be measured in terms of gross domestic product; employment in terms of the percentage of the labour force employed; agricultural, forestry and fisheries production in terms of crop, timber and fish yields and the income derived from them. While such easily measured production may be won at the cost of diminishing the resource base, and although conservation can bring real benefits by securing that resource base, the costs and benefits are not readily related[1].

Not all governments have explicit conservation policies; and the policies that exist tend to be narrowly sectoral. Consequently opportunities for the joint planning and realization of the conservation requirements of agriculture, forestry, fisheries, wildlife, and so on, may be overlooked. Indeed, the policies of the sectors concerned may conflict. Similarly, the interests of sectors not usually thought of as deriving benefits from living resource conservation may be neglected. Health is an example: conservation can advance the achievement of health objectives not only by ensuring a healthier environment—for example through the maintenance of clean air and water—but also by preserving genetic resources needed for the production of medicines. Policy makers in the health and industrial sectors need to be satisfied that the genetic resource-base of domestic pharmaceutical manufacture can be secured.

Action Required

Anticipatory Environmental Policies

Policies that attempt to anticipate significant economic, social and ecological events rather than simply react to them are becoming increasingly necessary for the achievement of several important policy goals: the satisfaction of basic needs, such as food, clothing, sanitation and shelter; the development of a high quality environment; the optimum use of available resources; and the control of pollution and other forms of environmental degradation. Achievement of these goals requires not merely policies that promote recycling, reduce the production, marketing and disposal of products dangerous to the environment, and make economic use of residual wastes. It also requires policies that actively promote human health and well-being, the protection of the living resource base, and the adoption of resource-conserving settlement patterns, transport systems and modes of trade and consumption. Such anticipatory environmental policies involve actions to ensure that conservation and other environmental requirements are taken fully into account at the earliest possible stage of any major decision likely to affect the environment. They are not intended to replace reactive or curative policies; simply to reinforce them.

Adoption of anticipatory environmental policies may pose difficulties. By their nature they require action before damage to the environment has created a demand for it. They also incur the costs of planning, research and preventive action and sometimes of delays or modifications to particular developments. Yet in general these difficulties are heavily outweighed by the advantages. Anticipatory policies enable societies to avoid the high and usually recurring costs of environmental mistakes—mistakes that can frustrate development objectives, waste resources, and impair the very capacity for development. Measures to prevent environmental degradation taken at the design stage of products and development projects alike are normally more cost-effective than measures taken once a problem has arisen when they may require redesign, restructuring, the banning of a product or the abandonment of a partially completed project. Anticipatory measures are often not only beneficial to society—avoiding high external economic, social and health costs—but also profitable to the enterprise concerned. To avoid delays environmental factors should be taken into account when a project is first formulated and subsequently in parallel with project development.

Cross-sectoral Conservation Policy

It is recommended that governments adopt a cross-sectoral conservation policy to:

— commit themselves to achieving the objectives of conservation;

— define the conservation requirements and responsibilities of the various government sectors in relation to those objectives;

— indicate a timetable or target dates for meeting the requirements and carrying out the responsibilities.

Depending on the constitutional structure of the country concerned and on the distribution of planning responsibility and powers of decision over resource use, conservation policies may be required at more than one level of government. The roles and responsibilities of each level of government should be made clear and effectively related to those of the levels of government above and below it. In addition, national policies should include specific guidance on the giving and receipt of aid and technical assistance, as well as on international agreements relevant to conservation. Several agencies may be responsible for preparing national positions with respect to international organizations and these should be harmonized: consistent positions should be developed for the governing bodies of United Nations and other intergovernmental organizations on which the governmentsits.

As a matter of policy, agencies re sponsible for living resources should be as much concerned with maintenance as with production. The need for food, fuel and fibre and other natural products, as well as for foreign exchange, may temptliving resource managers into encouraging or permitting overexploitation of the resources in question or the undermining of the ecological processes and genetic diversity on which they depend. This is highly likely if policy goals are concerned mainly with production and only incidentally with maintenance. The policies of agencies responsible for living resources should also require that each sector's conservation responsibilities are discharged with the conservation needs of other sectors in mind.

For example, the policy goals of forestry should include:

to increase yields of goods and services from forests, such as wood products, water, wildlife, recreation, education and research, provided that

— such yields are sustainable, and that

— the resource base (essential ecological processes and genetic diversity) is secured;

to manage the forest estate on the principle of stewardship, with commitment to maintain in perpetuity ecological processes, watersheds, soils and genetic diversity; to assist other sectors (such as agriculture, rural development) to make efficient, ecologically sound use of forest goods and services.

Similarly, the policy goals of agriculture should include:

— to supply food and other agricultural products in sufficient quantity and of acceptable quality, consistent with the maintenance of the resource base, particularly soils, water, the habitats of organisms necessary for pollination and integrated pest control, and the genetic diversity of crops, domestic animals and their wild relatives;

— to maintain and enhance the quality and attractiveness of rural areas; to recycle nutrients, ensuring that crop residues and livestock wastes are returned to the land, controlling pollution, and assisting where practicable in the recycling of urban wastes.

Integrating Conservation and Development

Conservation can be integrated with development through the instruments used to implement anticipatory environmental policies, through the establishment of coordinating mechanisms to ensure that a cross-sectoral conservation policy is applied, and by the adoption of national accounting systems to include measures of conservation performance. Instruments for the implementation of anticipatory environmental policies include: taxes, charges and financial incentives (to encourage choices compatible with the maintenance of a healthy environment); technology assessment; design and product regulation; environmental planning; and procedures for rational use allocation.

Although the deficiencies of gross domestic product as a measure of national welfare are generally acknowledged, national accounting systems and many policy decisions are still couched in strictly monetary terms. Consequently the costs of conservation and of measures to enhance human welfare in other ways may often appear to outweigh the benefits, since the costs are entirely calculable in money while the benefits are not. In order for governments to take-adequate account of the costs of destroying, degrading or depleting living resources and of the benefits of conserving them, it is recommended that nonmonetary indicators of conservation performance be selected for inclusion in national accounting systems. This is easier said than done, but suitable indicators might be:

a. extent of most suitable agricultural land that has not been lost to non-agricultural activities or degraded by poor farming practices;

b. silt load of rivers as a proportion of the size of the river basin (an index of erosion);

c. proportion of unique species and of unique varieties of domesticated plants and animals and their wild relatives whose survival is secured;

d. proportion of ecosystems and species that are being utilized sustainably.

This list is illustrative; but all are possible measures of the state of important national assets (good soil, secure catchment areas, and genetic and other renewable resources). Nevertheless, further consideration needs to be given to the selection of ecological indicators to ensure that their measurement and. monitoring is practicable and that in combination they provide a reasonable reflection of conservation performance.

Environmental Planning and Rational Use Allocation

The problems

Environmental planning and the allocation of uses on the basis of investigation and planning are essential if optimum use is to be made of available resources. Without them, the prospects of sustainable development will be impaired, sometimes permanently. For example, dams may be sited so that they drown and destroy highly productive land or important areas of genetic diversity. Pollution emission standards may be set so low that acid rain reduces the productivity of forests and fresh waters, or pathogens and heavy metals contaminate food (such as shellfish) rendering it unmarketable or, if it is marketed, directly damaging human health. Industries and settlements may be built on the best farmland or on land "reclaimed" from coastal wetlands, thus reducing the productivity of agriculture and fisheries.

Action Required

Ecosystem Evaluation

Every use of the land, fresh waters and the sea has its own site requirements, as well as different degrees of compatibility with other uses. Equally, every ecosystem has its own characteristics which make it more or less suitable for a particular use. The function of ecosystem evaluation is to assess the characteristics of ecosystems and match them to the most appropriate uses. Ecosystem evaluation is carried out in many guises under many names: land evaluation, land capability assessment, land suitability assessment, and so on. Some times a rather limited range of potential uses is considered: for example, agriculture, livestock production, wildlife production, forestry. Often only land areas are assessed. Here the term ecosystem evaluation (EE) is preferred to land evaluation for two reasons: to make clear that marine and freshwater areas should be evaluated as well as land areas; and that the areas being evaluated are dynamic ecosystems with linkages of varying strength to other ecosystems. A particular use of an ecosystem may be incompatible not only with.other possible uses of that ecosystem but also with certain uses of other ecosystems. For example, a coastal wetland might be assessed for its suitability as a nature reserve for birds (if protected), as a harbour (if dredged) or as farmland (if filled in); but the wetland might also be an important nursery area and source of nutrients for a valuable fishery elsewhere, with which only the first use would be compatible. Any evaluation would be incomplete if such interrelationships were not taken into account.

Certain principles are fundamental to the approach and methods employed in ecosystem evaluation[l]:

a. *Ecosystem suitability is assessed and* classified with respect to specified *kinds of use.* The concept of ecosystem suitability is meaningful only in terms of specific kinds of use, each with its own requirements. The characteristics of each ecosystem are compared with the requirements for each use.

b. *Evaluation requires a comparison of* the outputs obtained and the inputs *needed for each different use.* An ecosystem that is not being used may still be useful: for example, an unexploited tract of forest will moderate the local climate, regulate water flow, and so on. Conversely, even non-consumptive uses may require the expenditure of resources other than those present in the ecosystem concerned: a nature reserve preserving genetic diversity requires measures for its protection; a recreation area requires roads and other infrastructure. Suitability for each use is assessed by comparing the required inputs with the outputs obtained. The input-output comparison should be quantified only to the extent that quantification does not distort what is being compared and does not attempt to compare what is not comparable. While inputs (such as labour, money, fertilizers) can normally be expressed without distortion in money terms, many outputs cannot. Quantification, therefore, should be done with great care and any assumptions should be stated explicitly.

c. *An interdisciplinary approach is required.* The evaluation process requires the integration of contributions from ecology and related natural sciences, the technologies of ecosystem use (agriculture, forestry, fisheries, and so on), economics and sociology.

d. *Evaluation is in terms relevant to the* physical, economic and social context *of the area concerned.* Such factors as the regional climate, standards of living of the population, availability and cost of labour, need for employment, the local or export markets, systems of land tenure that are socially and politically acceptable, an'J availability of capital, form the context within which evaluation takes place. Assumptions about such factors should be stated explicitly.

e. *Suitability refers to use on a sustained basis.* The possibility of degradation and depletion should be taken into account when assessing suitability. For example, there might be forms of use highly profitable in the short run but likely to lead to soil erosion, progressive pasture degradation, or adverse changes in river regimes downstream. Most changes of use, other than protection, involve a loss of equilibrium in the ecosystem concerned, sometimes a radical loss, as when a forest is converted to farm land; but there should be a strong probability that the new equilibrium will be long-lasting. Essentially, this requires that the resultant impacts on ecological processes (such as soil regeneration), genetic diversity and the productivity of ecosystems and species should be assessed as accurately as possible.

f. *Evaluation involves comparison.* Comparison can be between an existing use and a potential use, between potential uses, or between a potential consumptive use and a non-consumptive use. Comparison can also be broad or fine: for example, between agriculture and forestry, between two or more farming systems, or between one crop and another.

Ecosystem evaluations should be prepared as a matter of priority and be used to influence all policies from the earliest possible date. An interim EE may be necessary in countries in the middle of a multi-year planning cycle—for example, a five year or ten year economic or development plan. The preparation of EEs should be based on available knowledge and should not be deferred because of lack of knowledge.

The degree of detail and area of coverage possible will vary greatly from country to country. It is preferable to achieve coverage at the expense of detail rather than the other way round. In due course the evaluation can be completed in greater detail area by area. Areas scheduled or proposed for development should be given priority.

Assessment of Environmental Effects

An assessment of environmental effects is an activity designed to identify, predict, interpret and communicate information about the effects of an action—be it a policy, programme, legislative proposal, engineering project or other operation with environmental implications—on human health and well-being, including the wellbeing of ecosystems on which human survival depends. Environmental assessments are a means of ensuring that ecological and social information is included with physical and economic information as the basis for making decisions.

Environmental assessments should be an integral part of the planning of all major actions (both public and private) requiring government authorization. They should be carried out at the same time as engineering, economic and socio-political assessments; and should examine alternatives to the proposed action. Major actions should be interpreted as including tax and fiscal policies, as well as other policies which by influencing investment can have a significant impact on the environment. It should be the responsibility of government agencies to assure the quality of environmental assessments. If an assessment is prepared by the proponent of an action—whether a private party or a government agency—there should be a mechanism for independent review. To avoid delays environmental assessments should be prepared as early as possible in decision making, from the preliminary study stage onwards, examining acceptable actions in increasing detail as their formulation proceeds. Experience suggests that the costs of environmental assessments vary considerably but need not be high: in the USA, for example, they range from 0.05% to 2% of the value of the project assessed.

A procedure for Allocating Uses

To make optimum use of available living resources, it is recommended that land and water uses be allocated as follows. First, uses should be allocated tentatively according to their compatibility with each ecosystem's capacity to supply particular goods and services (or fulfil particular functions). The ecosystem evaluation, elaborated as necessary by environmental assessments, will achieve this first step. Next, uses should be allocated tentatively a second time on the basis of current and projected patterns of demand on each ecosystem as reflected by current use. Here, demand equals present uses of, plus impacts on, ecosystems. Current uses of each ecosystem should be identified, and projected increases and changes in demand indicated. At this stage demand for non-living resources (construction materials, minerals, oil, gas, space for roads and buildings) as well as energy consumption and settlement patterns should be included.

Finally, the results of allocation by supply characteristics (the EE) and of allocation by demand characteristics should be compared to reveal conflicts and compatibilities between the two. In the case of compatible uses care should be taken to define precisely what those uses are and to ensure that management systems are available to manage the areas concerned on a multiple-use basis. Management requirements to ensure retention of those characteristics of the ecosystem that permit multiple-use should be noted. Conflicts should be reconciled where possible by zoning and scheduling. Where this is not possible their resolution will be a matter of political judgement. However, uses that depend on unique or irreplaceable ecosystem characteristics should have priority over other uses. For example, an ecosystem supporting a critical habitat of an endangered species should as far as possible be managed to protect that habitat.

The allocation procedure outlined is an integrating mechanism. It enables policy makers to confront simultaneous-ly ecological, social and economic cri-teria and therefore to make informed choices before resources are irrevocably committed. It can suggest those develop-ment opportunities likely to be both productive and sustainable, and show where trade-offs between one policy and another may be expected to be large or small. If all policies were adjusted at this point, many resource conflicts could be minimized, and others resolved without social or economic disruption.

Use allocations, together with eco-system evaluations, environmental as-sessments, and other supporting data and opinions, should be made available to the public so that the political process is properly informed. The public should be given adequate time and opportunity to consider relevant information and to influence decisions. Research needs should be identified concurrently with the preparation of ecosystem evalua-tions and use allocations. Use alloca-tions should be evolutionary in nature, being modified from time to time in the light of events, new knowledge, and changes in perceived needs, aspirations and values. The results of policies, eco-system evaluations, use allocations, and of other actions and decisions—whether or not they are directly concerned with conservation—should be monitored and regularly evaluated. *The assumptions behind all such actions and decisions should be stated explicitly so that they can be tested.*

Improving the Capacity to Manage

In addition to the integration of conservation and development, sound living resource management requires effective legislation, organization, training and information. It is recognized that governments differ greatly in their constitution, organization and responsibilities.

The Problems

The development of conservation law, like that of environmental law in general, until recently has been somewhat piecemeal and haphazard—in response to sectoral needs and, sometimes, to emergencies. Consequently, legislation concerning living resources in many countries is marred by gaps, duplication and even conflicts. A still more common and especially serious problem, however, is the failure to implement laws and regulations whatever their quality. Sometimes lack of implementation is due to the law being so stringent that people must flout it to survive. Generally, however, it is because the law implies a governmental commitment and infrastructure or a degree of public understanding and support that simply do not exist. Legislation may, for example, authorize the sale of pesticides only upon a written affidavit that the pesticide has been tested—yet the facilities for testing may be inadequate. Often budgets are entirely inadequate for enforcement, penalties are weak, and jurisdictional conflicts between agencies or between central government and local government prevent the law from being implemented.

Two common failings of organization are a lack of coordination among agencies responsible for living resources, and mandates that charge the same agency with both the exploitation and the protection of a resource. Living resources are commonly the responsibility of several different agencies: for example, agriculture, forestry, fisheries, wildlife, rural development, water. The consequences of such fragmentation vary. On the one hand, conservation arguments may be diluted and positions weakened—especially if the views of conservation units are not reflected in the official positions of the larger administrations of which they are part. On the other hand, the consequences may be positive: the presence of a conservation administration within a number of different government departments means that conservation arguments and positions may be brought to bear in many more policy areas than they would be if lumped in a single department. This is more likely to be the case if the mandates of each department specify conservation. It is therefore not essential to combine these and closely related agencies such as public health in departments of environment and natural resources, but it is necessary to provide a coordinating mechanism for living resource agencies in order to bridge gaps and reduce conflicts and duplication. Similarly, while the combination of use and protection of a resource is perfectly rational, the mandates of agencies charged with the two duties should specify conservation, making it clear that unsustainable exploitation of the resource concerned would be a breach of the mandate.

The need for cross-sectoral coordination is particularly important in the cases of soil conservation and the conservation of marine living resources. Few developing countries have the organizational capacity to check the rapidly increasing loss of land and water resources by erosion and to prevent further loss through effective soil and water conservation measures. Still fewer countries (whether developing or developed) are able to manage efficiently their use of the sea. Like the land, the sea is an area of multiple use. It is used for food production, transport, mining and quarrying, oil production, recreation and waste disposal. Unlike the land, however, very little attempt is made to manage for multiple use. Such regulatory bodies as exist are generally concerned with a single resource, such as fish. As a result, the seas and their living resources are increasingly being overex-ploited and degraded.

The capacity to manage living resource use can also be weakened by the division of conservation responsibilities among different levels of government—normally central (or national), provincial (or state) and local (or municipal). Since ecosystems and species seldom observe institutional or political boundaries it is most important that there be close coordination among these levels. It is also most important that the jurisdictions of the different levels of government be clearly defined. Conservation opportunities can be missed and problems left unresolved when disputes over which government is responsible for what cannot be settled.

Another widespread problem, especially in developing countries, is the lack of skilled personnel. Often this is due to inadequate training facilities; but low salaries (in comparison with the private sector) and poor administrative organization are also significant factors. As an example of this last, the few trained foresters available may be located only in national and provincial capitals. In Thailand, because field staff are paid less than headquarters staff, and because of the hardship of life in the field, more than half the trained foresters work in the capital'.

Action Required

Legislation

Each country should review and consolidate its legislation concerning living resources to ensure that it provides sufficiently for conservation. Each country should also review—and if necessary strengthen—its capacity to implement its conservation legislation, both existing and required. Ideally, a commitment to conserve the country's living resources should be incorporated in the constitution or other appropriate legal instrument. The commitment should lay down the obligation of the state to conserve living resources and the systems of which they are part, the rights of citizens to a stable and diversified environment, and the corresponding obligations of citizens to such an environment.

There should be specific legislation aimed at achieving the objectives of conservation by providing for both the sustainable utilization and the protection of living resources and of their support systems. Comprehensive conservation legislation should provide for the planning of land and water uses and should regulate both direct impacts on the resource, such as exploitation and habitat removal, and indirect ones, such as pollution or introduction of exotic species. In addition, it should include requirements to undertake ecosystem evaluations, environmental assessments, and like mechanisms to ensure the incorporation of ecological considerations into policy making. The law should also provide for the participation of citizens in the elaboration of policies, for the provision of sufficient information for participation to be effective, and for legal recourse to implement these rights. In addition there is a need to revise traditional concepts of the law of remedy, which currently envisage compensation only for economic loss, narrowly defined, and do not provide for indirect or long term damage to individuals and communities through the depletion of species or the destruction or degradation of ecosystems.

Special attention should be paid to the enforcement of conservation law. Enforcement is a multidisciplinary activity that should begin with the design of legislation. It is necessary but not sufficient to provide adequately trained and funded personnel to implement and police the law. It is also important to make sure in advance that the law is ecologically, economically and socially feasible. Public education programmes may be required both before and after the law comes into force to help the public understand and support it. If the law imposes undue hardship on a particular segment of society, then measures to relieve that hardship may be needed. The effects and the effectiveness of the law should be monitored so that, if necessary, the law or its enforcement can be improved.

Organization

Governments should review the status, organization and funding of agencies with responsibilities for living resources. They should take the necessary steps—including changes in legislation—to ensure that conservation policies are implemented and that the agencies concerned have the resources and the staff to carry out promptly and fully ecosystem evaluations, environmental assessments and any other measure required for the conservation of living resources.

The following principles should form the basis of organization within government to achieve conservation:

a. the different agencies with responsibilities for living resources should have clear mandates and such man dates should specifically include conservation;

b. there should be a permanent mechanism for joint consultation on and coordination of both the formulation and the implementation of policies;

c. such a mechanism can be achieved by giving new authority to existing agencies or by establishing new units in existing agencies; by setting up comprehensive agencies responsible for all living resources; or by setting up cabinet-level units to ensure that all sectors concerned carry out their conservation responsibilities;

d. each agency should be required by statute to disclose and explain its positions to the public;

e. policies and decisions should be implemented; sufficient financial and other resources should be provided to make this possible.

The more limited the availability of trained planners and managers the more important it is to avoid dispersing them among agencies with narrow mandates and conflicting aims. To encourage recruitment at the technical level it may be necessary to provide professional recognition to technicians.,Where disparities between private sector and public sector salaries increase the shortage of trained personnel, public sector salaries should be increased. Similarly, the salaries of field personnel should be at least as high as those of headquarters staff— indeed they may need to be higher to compensate for poor conditions.

Because soil and water conservation involves environmental planning and many kinds of land use (for example, agriculture, livestock raising, forestry, mining, road building) a special organization is needed to promote and coordinate conservation measures. It is therefore recommended that a soil and water conservation body be established at the policy making level. If no related organization exists, a high level technical unit should be set up in association with the policy making body to initiate and coordinate operations. If soil or water management services already exist within some agencies, they should be brought into a single unit, combining: aspects of land use and of soil and farm management relevant to conservation, erosion control, soil conservation, land consolidation, range management, irrigation and drainage, flood control, surface water storage, groundwater resources, and so on. Since a full-scale unit, as outlined here, would initially be beyond the reach of many countries, it would be sensible to form a small comprehensive watershed planning service as a first step. This service could start off with relatively small projects covering a few lesser watersheds, allowing it to acquire expertise and accumulate greater responsibilities over a period of time.

New organizations are needed—or mechanisms need to be established to coordinate existing organizations—so that marine living resources can be managed comprehensively rather than along sectoral lines. The basic unit of management should be the ecosystem. There should be close cooperation among the organization or organizations responsible for the living resources of one ecosystem and those responsible for ecosystems linked to that ecosystem by significant exchanges of nutrients or movements of species. There should also be close cooperation between marine management bodies and the authorities responsible for contiguous areas of the land, since impacts on freshwater systems and on coastal habitats greatly affect marine living resources. Coordination should be especially close between organizations responsible for fixing catch levels, closed seasons and other fisheries regulations and those charged with ensuring the integrity of the habitats on which the marine living resources depend. Great care should be taken to ensure that coastal wetlands, shallows and other critical marine habitats are properly protected from pollution and other forms of inappropriate modification. Full account should be taken of the effects of changes in ecosystems on the species utilized and of changes in catch levels on ecosystems.

Training and Research

The Problems

A major constraint on the implementation of conservation measures is a lack of trained personnel. In many countries the lack of environmental lawyers, for example, means that out-of-date laws are not revised or another country's legislation is duplicated without being adapted to local conditions. There is often an acute need for people trained in living resource management, such as foresters, and watershed managers. Indonesia, for example, currently has only 400 foresters, or one forester per 3000 km of forest (or fewer than one per timber concession). The list of scientists and professionals needed by developing countries is long: ecologists, geologists, hydrologists, public health engineers, environmental economists, environmental planners, and so on. Even where professional staff are available there is an acute shortage of technicians: scientists, for example, may find themselves having to maintain their own equipment. Sometimes the shortage of technicians is exacerbated by the shortage of professionals, because successful trainee technicians may decide to continue their education so that they can achieve the higher status and salaries of the professions.

Many countries also lack adequate information. Generally this is because the countries' data gathering capabilities are weak, but even when they are satisfactory information flow is hampered by poor data retrieval and distribution systems. As a result of such deficiencies countries lack the information base necessary for rational resource planning and management: for example, the extent of forest cover and the rate of its removal; aquatic pollution levels and assimilative capacities; and species inventories for protected areas. Comprehensive air and water monitoring systems are so expensive and sophisticated that only developed countries can afford them; but not enough is known about the dynamics of tropical ecosystems to develop less expensive but equally reliable systems using indicator species. The level of applied research on ecosystems and their modification needs to be stepped up considerably if policy makers are to be given better advice on such matters as the extent to which coastal wetlands can be modified, the pollution absorption capacities of fresh waters, and the most favorable cropping patterns for integrated pest control.

Although a great deal is known about many species and ecosystems, what we know about the biosphere is less than what we do not know. The dynamics of many important ecosystems and the relationships among ecosystems are also poorly known. It is therefore seldom possible to predict accurately the effects of human actions on a great many ecosystems—at least not in a way that might be useful to a policy maker— without special and often lengthy research. The same generalization applies to determining sustainable yields from multi-species fisheries. Such lack of knowledge often causes difficulties between policy makers and resource managers on the one hand and the ecologists and other scientists that advise them on the other. The former expect a clarity and precision of advice that is premature (and, if attempted, may well make the advice wrong); the latter cannot avoid stressing the real and important uncertainties that exist.

Governments and resource users are scarcely ever in a position to defer action pending the outcome of a protracted research programme. Yet action based on inadequate knowledge carries a grave risk that it will fail or be unnecessarily destructive. Unacceptable consequences of lack of knowledge are best avoided (as far as possible) by good planning and management, so that development activities can be so located and conducted that risk is reduced. At the same time management needs to be more research-oriented and research more management-oriented so that the most urgently required knowledge is generated most quickly.

Action required

Training

Each country should review the capacities of its universities and other centres of higher education to train professionals and technicians in the expertise and skills necessary for planning and managing the use of living resources. National and regional training facilities should be strengthened as appropriate. Training is required at three levels: professional ; technician; user.

At the professional level there is a need both for specialists (individuals able to make detailed studies, surveys, and designs for specific practices) and for generalists (individuals with a broad grasp of the theory and practice of conservation—either within a sector or cross-sectorally—and with an overall understanding of the various disciplines involved). Appropriate university or college courses are required to meet either need. The interdisciplinary courses required for the training of generalists may well have to involve several faculties—as indeed will the training of some specialists, such as soil conservationists, where the subject demands knowledge of a number of disciplines (such as, in the case of soil conservation for example, agriculture; range management, forestry, civil engineering, hydrology, ecology). Research components of university courses should be relevant to the student's own country and preferably should be carried out there. Universities providing courses for foreign students should make every effort to provide for this requirement.

At the technician level there is an acute need for people trained to operate in the field as agricultural and fisheries extension officers, wildlife and protected area managers, soil conservation workers, foresters, and so on. This need is probably most effectively met by a combination of institutional and in-service training. Institutional (post-secondary school) training both enables the student to enhance his or her basic educational skills (where necessary) and provides recognition of status through the conferral of diplomas. In-service training enables the student to acquire rapidly essential practical experience. To help countries build up cadres of trained personnel, overseas organizations operating in developing countries should include counterpart training in every project.

Finally at the user level, farmers, pastoralists, fishermen, loggers, plantation operators and other land and water users need to be trained in production methods that are both sustainable and more productive in the long term than present practice. This requires that extension services be staffed with sufficient numbers of extension workers to maintain effective contact with land/water users and of specialists to provide extension workers with adequate technical support. To be successful, extension services must take great pains to explain to land/water users the need, purpose and expected results of any measures they recommend. Demonstration, normally involving the more responsive members of the community, will be most important for convincing the community at large of the value of such measures.

Research

Although there will always be a need for more knowledge, it is most important that the considerable body of knowledge that already exists be used. Too often the need for additional study is put forward as an excuse for not taking conservation action. By contrast, development projects are too often initiated without sufficient study of their potential impacts. Environmental assessment of development projects and other actions may reveal the need for research. Whether such actions should be deferred pending the outcome of the required research will depend on the circumstances of each case. Prior relevant research can reduce the necessity for such hard choices.

Governments should place living resource research high in their national scientific and research programmes. They should establish national councils to encourage universities and other bodies to increase and coordinate their living resource research activities and to relate research to conservation action on the ground. Research programmes should cover three broad overlapping areas:

inventory—this includes research on the distribution of ecosystems and species in each country;

functional—this includes research on ecosystem dynamics and relationships, the effects of human activities on ecological processes and vice versa, baseline monitoring, and other basic ecosystem, species and population studies;

management-oriented—this includes research into standards, techniques and technologies that will improve the planning and management of living resource use.

Inventories and functional studies provide essential information for ecosystem evaluations, for decisions on the optimum distribution and management objectives for protected areas, for decisions on where particular conservation measures are most needed, and for many other aspects of planning and management. Important studies falling under these headings include: field studies, including mapping, of the location, extent and severity of erosion; mapping of the relationships between the distribution of. important, rare or threatened species, ecosystems and their support systems, and of actually or potentially damaging human activities; research on those social and institutional factors contributing to living resource problems, contributing to the solution of such problems, or acting as obstacles to solutions.

Although such studies are not so directly management-oriented as are, for example, investigations of the productive capacity of a fishery so that a catch level can be set, or of the assimilative capacity of a river so that a pollution standard can be fixed, they make a substantial contribution to management effectiveness. Indeed management in the broadest sense can be regarded as including assessment, research and monitoring.

Accordingly managers should be concerned as much with generating new knowledge to improve management of the resources for which they are responsible as they are with implementing management decisions made in response to current knowledge. Often this will mean increased emphasis on experimental management, for example by deliberately overexploiting one stock and un-derexploiting another to test whether otherwise untestable assumptions about the state of the exploited population are valid. Experimental management is likely to be the most rapid and reliable, and frequently the only way of determining what production systems (whether of agriculture, livestock production, forestry, fisheries, or combinations of these) are both highly productive and sustainable.

A non-experimental, passive approach to management, by contrast, may lead to the adoption of systems that are either unnecessarily conservative (with the desire for sustainability leading to a loss of output) or recklessly exploitative (with the desire for output leading to collapse of the resource).

International Research

Much research necessary for the management of living resources can be conducted most cost-effectively either by international organizations or by national organizations working within internationally coordinated research programmes. The latter are particularly useful, not only for the study of large-scale phenomena such as climate and biogeochemical cycles but also as a means of avoiding unnecessary duplication of research effort. Examples of such programmes include Unesco's Man and the Biosphere (MAB) Programme and the work of the Scientific Committee on Problems of the Environment (SCOPE) of the International Council of Scientific Unions (ICSU). MAB provides a valuable opportunity for an integrated research programme into ecosystems and ecological processes, using the world-wide network of biosphere reserves, to provide a strategic ecological information base. SCOPE provides, amongst other things, an international research agenda concerning essential ecological processes.

Participation and Education

Ultimately the behaviour of entire societies towards the biosphere must be transformed if the achievement of conservation objectives is to be assured. A new ethic, embracing plants and animals as well as people, is required for human societies to live in harmony with the natural world on which they depend for survival and wellbeing. The long term task of environmental education is to foster or reinforce attitudes and behaviour compatible with this new ethic.

The Problems

Lack of awareness of the benefits of conservation and of its relevance to everyday concerns prevents policy makers, development practitioners and the general public from seeing the urgent need to achieve conservation objectives. Ultimately, ecosystems and species are being destroyed because people do not see that it is in their interests not to destroy them. The benefits from natural ecosystems and their component plants and animals are regarded by all but a few as trivial and dispensable compared with the benefits from those activities that entail their destruction or degradation. Until people understand *why* they should safeguard ecosystems and species they will not do so.

There are two distinct problems: Public participation in conservation! development decisions is seldom ade*quate.* Consequently the decisions may not reflect sufficiently the experience and wishes of the people affected, and the benefits of the programme or project may be fewer than expected.

Although there has been progress, there is insufficient environmental education. Informal education programmes, directed at the adult public, are haphazard; and formal programmes, directed at schoolchildren and students, are still too few and inadequate. Relatively little can be achieved and few achievements will last while the contribution of conservation objectives to development and the requirements of achieving those objectives remain poorly communicated. Despite the enormous growth of conservation literature, there are few information materials designed to persuade people of the contribution of conservation to development or the relevance of conservation to the concerns of, say, business people, trade unionists or health officials. There is a wealth of emotional appeals directed at affluent audiences and of didactic explanations of how ecosystems work. But the cases for the maintenance of ecological life-support systems and for the preservation of genetic diversity are too often made anecdotally and without sufficient documentation to convince the sceptical. Furthermore, they have not been documented fully enough or described with sufficient precision to provide guidance to policy makers on what ecological processes and genetic resources are indispensable and should be secured as a matter of priority.

Action Required

Public Participation

Local community involvement and consultation and other forms of public participation in planning, decision making and management are valuable means of testing and integrating economic, social and ecological objectives. They also provide a safeguard against poorly considered decisions and an indispensable means of educating both the public in the importance and problems of conservation, and policy makers, planners and managers in the concerns of the public. Participation tends to build public confidence and improve the public's understanding of management objectives. It provides additional data for planners and policy makers. Public participation is particularly important in rural development, for without the active involvement of the people—including identification by them of the problems that most need tackling and how to deal with them—little can be achieved.

The extent of public involvement in the development planning process depends on both the attitude of the government and the interest of the community. Ideally, however, public participation should be at all stages of the development process from policy making to project formulation and review.

At whatever stage it is involved, the public should be given time and information sufficient for it to influence decisions.

Benefits of Public Participation in Rural Areas

More information is acquired about local needs, problems, capabilities and experience; effective planning and implementation require specific information of the sort only local people can provide efficiently. Better plans can be made that are more realistic about what is possible, what will be done, and what suits actual conditions best. To the extent that local people are involved in decision making and have had some voice in what is to be done with local resources, they may contribute labour and funds, as well as land and materials. Implementation will be smoother and quicker once understanding and assent have been generated through participation—as people usually cooperate more willingly in decisions in which they have participated.

Talent for management and administration can be developed in the rural sector to complement that of the government.

Integration of activities and services will be more effective and complete. Maintenance of investments in roads. canals, terraces, buildings and other facilities is usually better where the local people have been consulted and involved in their creation. Political support will be greater where the facilities and services created under government auspices are those identified by rural people as more important and valuable. The judgment of the people affected by development programmes is essential for the evaluation of such programmes. Source: 1.

Environmental Education Campaigns and Programmes

If the users of living resources (farmers, fishermen, foresters, industries based on living resources, recreational users, and so on) are unaware of the need to conserve the resources they are using, an education campaign should be prepared for them; the same goes for other groups that may have an impact on living resources, even if they do not use them so directly, if they are unaware of the need to manage their activities in ways that are as compatible as possible with conservation. If, however, government does not recognize the need to meet the conservation requirements concerned, special efforts will be needed to direct information on the importance of such requirements to the appropriate legislators and decision makers.

Advantage should be taken of circumstances when pro-conservation decisions are evidently the most profitable within the time-frame of concern of legislators and decision makers, namely: when the leaders are personally convinced that conservation policies are the right course to pursue; when the electorate is so convinced and makes it clear that it will vote for those policies; when influential groups within the country are educated in and committed to conservation policies; when pro-conservation decisions are an effective way of achieving other policy objectives.

Organizers of education programmes should determine the main target groups of such programmes, define precise programme objectives, and select the media and techniques that are most effective with the target groups. Results, together with the techniques and materials used, should be regularly evaluated against the stated objectives. The most important target groups are: legislators and administrators; development practitioners, industry and commerce, and trade unions; professional bodies and special interest groups; communities most affected by conservation projects; schoolchildren and students.

Education programmes directed at any of the first three groups require clear, succinct information materials showing the contribution of conservation to the achievement of the policies and goals that most concern the target audience. There is a particular need for well documented accounts of the extent and manner in which societies at different stages of development depend on the achievement of each of the three objectives of conservation. Also needed are collections of case histories of sucessful conservation and sustainable development projects.

Whenever possible, education programmes should be included in all conservation and resource-use projects in order to improve local understanding and support for conservation and to enhance the projects' prospects of lasting success. These programmes should supplement programmes to train and equip local communities in improved resource-use practices. Existing agricultural and other extension services should be used to promote conservation; and literacy programmes should include conservation material.

School curricula should include environmental education both as an intrinsic part of other subjects (so that conservation attitudes can influence all activities) and as a separate subject (so that ecology can be taught more formally and its concepts more readily grasped). Inexpensive teaching materials (textbooks, audio-visual aids, posters, pamphlets, and so on) should be prepared. The materials should explain ecological concepts and the objectives of conservation, using local examples wherever possible. The effectiveness of teaching materials should be regularly evaluated. Environmental education should also be an important part of the out-of-school activities of children. Wildlife clubs should be encouraged; and environmental education included in the activities of youth groups.

As well as focussing on special interest groups, the several mass media (radio, television, newspapers and periodicals) should be fully used to reach the general public. Advantage should be taken of those occasions when the public comes into contact with plants and animals—in national parks and other protected areas, in zoos and botanical gardens, and in natural history museums—to explain conservation objectives and their contribution to human survival and wellbeing. In addition, special areas should be set aside for training, demonstration and education in ecology and conservation (for use by schools, universities and the general public). Such conservation education areas, besides serving their essential educational function, could also help take pressure off reserves protecting particularly fragile or unique ecosystems. Public concern for popular animals (such as whales) should be used to foster better understanding of the ecosystems of which those animals are part, and of how people both affect those ecosystems and benefit from them.

Certain living resource issues need much greater public exposure. As a matter of priority, an intensive public education campaign, explaining the effects of introduced species, should be directed at the pet trade and consumers, recreational hunting and fishing groups, governmental agricultural, wildlife and fisheries administrations, and at other bodies responsible either for allowing or for promoting introductions. Also understanding of and support for sustainable exploitation should be built up among both users and consumers of living resources; this is especially necessary with respect to those resources that are exploited commercially.

The need for environmental education is continuous because each new generation needs to learn for itself the importance of conservation. As such, individual campaigns and programmes should not be regarded as ends in themselves but as part of a long term, iterative process. It should also be recognized that any educational campaign is in competition for public attention with many others, including advertising campaigns. To win and retain as much of this attention as possible, it is essential for conservation to be seen as central to human interests and aspirations. At the same time, people—from heads of state to the members of rural communities— will most readily be brought to demand conservation if they themselves recognize the contribution of conservation to the achievement of their needs, as perceived by them, and the solution of their problems, as perceived by them.

Conservation-based Rural Development

Probably the most serious conservation problem faced by developing countries is the lack of rural development. In their struggle for food and fuel growing numbers of desperately poor people find themselves with little choice but to strip large areas of vegetation until the soil it self is washed or blown away. Often the rural communities responsible for this destruction do not need to be told it is a mistake: they are made acutely aware of it by an increasing lack of food, fuel, and other necessities. What such communities need is to be equipped to win their livelihoods in sustainable ways.

The Problems

More than 2,000 million people (about half the world's population) live in the rural areas of developing countries, and despite urban migration this number is expected to grow to almost 2,900 million by the end of the century. Most rural people live by farming, fishing and forestry or closely related activities.

Many are extremely poor, some 1,200 million people being classified by the United Nations as "seriously poor" (of whom almost 800 million are "destitute") with 500 million suffering from malnutrition. In their effort to satisfy their needs for food and fuel, the rural poor strip the land of trees and shrubs for firewood, clear steep and unstable slopes for cultivation, overgraze pastures, and overhunt and overfish the local wildlife. As a result, the daily survival decisions of the poor and hungry disrupt their own life-support systems, impair ecological processes and destroy genetic and other renewable resources just as surely as do too many of the development decisions of the rich and powerful in government and industry.

It is extraordinarily difficult for governments to deal with these problems because of their huge scale, because there are so many people and production units with which to deal—millions of villages and hundreds of millions of households and small farms—and because of the pace of change. Developments such as improved health services, better veterinary services, new wells, and higher yielding crop varieties, bring further changes to a situation that is already changing rapidly due to sheer pressure of numbers. Often, because they come separately and not as part of a coordinated rural development programme, such developments ultimately exacerbate the problems of the rural poor. For example, improved veterinary care, new wells, and the opening up of previously uninhabitable land by the eradication or control of diseases such as trypanosomiasis have enabled pasto-ralists to increase their livestock numbers and have provided them with new areas of grazing land in part compensation for areas lost to farmers. However, when these welcome developments have not been accompanied by effective provisions for better pasture management— and they seldom have been— the eventual result is usually heavy overgrazing and often irreversible soil degradation. Similarly, the change from shifting cultivation to settled arable farming—essential when the cultivation/fallow cycle becomes unstable and. pressure on soil and vegetation increases—can cause still greater erosion unless farmers are equipped to apply the necessary soil conservation measures. In rural development, as in development generally, the narrow sectoral approach is almost invariably self-defeating.

Unlike urban communities, the rural poor are dispersed over very wide areas. Therefore, rural people are less advantageously placed than their urban compatriots to bring their problems to the attention of government. For the same reason, their problems are less amenable to the kinds of development that governments with a narrow tax-base, inadequate institutions, poor delivery of services to rural areas, and a vociferous urban population, usually initiate. It is ostensibly easier, and certainly more visible, to plan, finance and manage a few large-scale projects—such as a colonization scheme, or a dam +irrigation + hydroelectricity—than to promote and oversee many village-scale projects. Yet the planning and management of most large-scale projects leave much to be desired, they are often short lived or marred by harmful side-effects, and they yield few benefits to the rural poor.

Action Required

There is an urgent need for rural development that combines short term measures to ensure human survival with long term measures to safeguard the resource base and improve the quality of life. A widespread requirement is the restoration of overexploited and heavily degraded living resources. Unfortunately, many rural communities are so poor that they lack the economic flexibility that would enable them to defer consumption of the resources in need of restoration. Conservation measures that require deferral of consumption will need to be complemented by measures that at least will maintain the rural comunity's standard of living and preferably will improve it.

Restoration and Compensation

If soil and vegetation need to be restored, they must be given a respite from intensive use. This requires integrated actions. In dryland areas these might include: the reduction of livestock numbers (possibly through price supports that encourage sale to market); increasing the efficiency of food production on nearby irrigated and rainfed farms; the employment of local people in replanting and reseeding schemes; and the provision of alternative settlement areas and alternative sources of water, fuel, food and other services (health, education, job training, and so on).

The full understanding and participation of local people in the formulation and implementation of these plans is crucial for success. The problem of persuading local people to participate is made easier if they can be shown the successful results of earlier projects. For this reason, the areas most likely to respond to rehabilitation should be given priority and used as demonstration projects.

The protection of a large number of relatively small areas is especially valuable in drylands to reveal what species are present, to provide for seed production and to demonstrate ecological recovery. Demonstrations of the benefits of restoring the full natural cover and productivity of the vegetation may be necessary to persuade local people that a relaxation of pressure on grazing lands is essential. Protected areas and other conservation measures, however, may restrict access to fuel, food, forage and other products. If substitutes are not then provided to compensate for the loss of these resources, the local community is likely to frustrate the conservation measures concerned.

Compensatory measures might include pasture improvement, the establishment of fuel-wood plantations, and the provision of credit or alternative food, fuel or fibre, as appropriate. If the measures concerned take time to bear fruit they must be supplemented by measures bringing immediate benefits. For example, if a protected area or a watershed forest is threatened by wood-cutting for fuel, it will be necessary not only to establish a fuelwood plantation but also to provide an alternative source of fuel that can be used at once. It would also be prudent to provide the community concerned with the means of conserving fuel supplies, such as more efficient cookers.

A rural community and a protected area may be entirely compatible, depending on the community's relationship with the ecosystems concerned and on the protected area's function. Many protected areas benefit local people directly by assuring a regular supply of water or by providing habitat for wildlife which may be hunted outside the protected area; examples are the Royal Chitwan National Park in Nepal and the Amboseli National Park in Kenya. However, the community should also be able to share in new benefits, such as income from recreation and tourism.

Although local communities may benefit indirectly from tourism if receipts by the national treasury are spent on such services as roads, water supply and health facilities, local commitment to a protected area can only be assured through provision of local advantages such as increased opportunities for employment and commerce.

Similarly, to foster local support and understanding, the local community should be involved in the protected area from the start—by being adequately consulted on management and by being given any employment or other economic opportunities associated with the protected area. Every effort should also be made to explain why areas are being protected and what the short and long term advantages to the local community are likely to be. Any feature of the protected area that may make it unique to the community or nation should be stressed, and people encouraged to regard the area as a source of local or national prestige.

For conservation-based rural development to be successful, there will need to be more research into sustainable systems of producing food and other goods from the rural sector, as well as training and incentive programmes to encourage and equip rural communities to adopt those systems known to work now. Among possible incentives are credit and marketing services adapted to the needs of the small farmer and fisherman and tax reductions or exemptions. An indispensable incentive and ultimately the one most likely to work is demonstration within the communities concerned that the new systems provide a higher quality of life using the resources at hand.

To provide demonstrations and training there needs to be a rapid strengthening of extension services. It is particularly important that extension services be interdisciplinary, multisectoral, and sensitive to the social and cultural characteristics of the area. Advice to farmers on how to improve yields or assistance to pastoralists to maintain better herds should be provided within an ecological framework with particular emphasis on soil and water conservation.

Traditional Knowledge

Rural communities often have profound and detailed knowledge of the ecosystems and species with which they are in contact and effective ways of ensuring they are used sustainably. Even when a community is growing in numbers and is clearly destroying a part of its environment it should not be assumed that all of this knowledge has disappeared or become invalid or that the traditional ways of regulating use have atrophied. Such assumptions tend to be self-fulfilling—with the result that a subsistence society with a prospect of developing harmoniously with its environment is more readily transformed into a poor cash society, hostile to attempts to regulate use and predisposed to degrade the environment still more severely.

Many traditional methods of living resource management are worth retaining or reviving, either in their original or in modified forms. For example, field experiments with traditional cropping systems in various parts of the world have demonstrated that many of these systems bring high yields, conserve nutrients and moisture, and suppress pests. The efficiency of traditional cropping systems can often be increased not by introducing completely different ones but by identifying those elements which could be improved and making the appropriate improvement. For example, Indonesian combinations of corn and rice have been shown to be both more resistant to pests and more responsive to applications of nitrogen fertilizer than are monocultures. The original strategy of the so-called "Green Revolution" of replacing tropical polycultures (growing a number of crops together) with temperate-style monocultures is itself increasingly being replaced by a new strategy: to retain the most productive elements of tropical polycultures and improve the remainder.

Priorities for International Action

Law and Assistance

Some living resources can be conserved only by international action. The principle of permanent sovereignty over natural resources together with recent extensions of national jurisdiction imply that the prime responsibility for conservation achievement lies at the national level. However, application of the principle is limited by the nature of many of the problems of conservation: many living resources are shared; many living resources occur (temporarily or permanently) in areas beyond national jurisdiction; living resources in one state may be affected by activities carried out in another.

Such resources can be conserved only by international action. International action is also necessary: to promote the conservation of genetic and other resources that are vital for the survival and wellbeing of all humanity; to stimulate and support national action.

It is recommended that cooperative programmes combining legislation, assistance and other actions be developed, concentrating on: tropical forests and drylands; the establishment of protected areas for the preservation of genetic resources; the global commons (the open ocean, the atmosphere, and Antarctica); regional strategies for international river basins and seas. These programmes should provide an essential focus for international action on those issues for which it is indispensable, as well as for international support for national action to implement the World Conservation Strategy.

International Law

Perhaps the most important form of international action is the development of international conservation law and of the means to implement it. Strong international conventions or agreements provide a legally binding means of ensuring the conservation of those living resources that cannot be conserved by national legislation alone. Because international conventions constitute a self-imposed restraint on each state's freedom of action, it is often difficult to conclude agreements of sufficient strength. Weak conventions, however, are dangerous and to be avoided, because they permit the illusion that problems are. being tackled when in fact they are not.

Although there are more than 40 multilateral conventions directly dealing with.the management of living resources, and although most have conservation components, few have conservation as their primary purpose. All international agreements relevant to living resources need to be reviewed to identify the most significant gaps in coverage, indicating whether each gap would best be filled by a global or a regional approach. The review should examine:

— *those conventions dealing directly with the management of living resources,* to determine whether they provide sufficiently for conservation;

— *those conventions whose primary purpose is conservation,* to determine whether they achieve the intended objectives and if not what measures are necessary to enable them to do so; preliminary suggestions with respect to the four main global conservation conventions follow.

Wetlands Convention

The Wetlands Convention at present lacks force, requiring only that states select at least one wetland for conservation, but not providing criteria to guide selection, guidelines for management, or adequate safeguards against delisting a wetland once selected. Experience has shown that an international convention must have a secretariat and a financial mechanism to be effective, but the Wetlands Convention lacks both. The convention could if it were revised become an important force for the protection of coastal and other wetlands essential for supporting fisheries (as well as for supporting water fowl, which is the chief concern of the convention at present); but in either case criteria for selection of wetlands and clear obligations on states for their conservation are essential.

World Heritage Convention

This convention recognizes the obligation of all nations to protect those unique natural and cultural areas which are of such international value that they are part of the heritage of all mankind; and the corresponding obligation of the international community to help them. It is important not only that states become party to the World Heritage Convention but also that they contribute generously to the World Heritage Fund. The Fund does not diminish the responsibility of each state to protect its unique natural areas, but it does provide a means of ensuring that areas of global significance are not lost because of a local lack of money or skills.

CITES

This convention has considerable potential, the key to which is its network of national management authorities and scientific authorities operating in direct communication with each other and the secretariat. The management authority is concerned with the mechanics of trade (such as permits); the scientific authority is concerned with the biological aspects and must therefore satisfy itself that the issuance of an import or export permit will not have a harmful effect on the species concerned. Developing countries often lack the personnel and financial resources to establish and maintain fully operational agencies for this purpose. Multilateral and bilateral development assistance agencies are recommended to provide assistance on request and to assist exchange of administrative and scientific experience among trading countries.

To make full use of the regulatory network established by CITES, certain improvements in its international standard-setting and national enforcement mechanisms are needed. At present, marine fishes, molluscs and crustaceans are inadequately represented in the CITES Appendices, and a further review of these groups is necessary. In addition, the administration of CITES at the national level needs to be linked much more closely with the existing customs, veterinary and phytosanitary inspection systems. Rather than inventing new bureaucracies, optimum use should be made of the available ones. Exploitation levels, trade, and response to regulations should be closely monitored, not only by governments but also by nongovernmental conservation organizations. Conservation organizations should monitor the implementation of CITES by their national (scientific and management) authorities. They should monitor trading in shops and through newspaper and other advertisements; and they should ensure that the annual reports and any proposals submitted by national authorities to the CITES secretariat properly reflect conditions, informing the CITES secretariat should this not be the case. Conservation organizations with monitoring experience could provide a useful service to other organizations by helping them to set up their own monitoring systems.

Migratory Species Convention

Migratory species, as defined by the Migratory Species Convention, are "the entire population of any species or lower taxon [such as a subspecies] of wild animals, a significant proportion of whose members cyclically and predictably cross one or more national jurisdictional boundaries". The Migratory Species Convention obliges parties to protect endangered migratory species and to endeavour to conclude agreements for theconservation of migratory species whose status is "unfavourable". Existing regional and bilateral agreements on migratory species have demonstrated that international conventions are the only effective means of protecting animals that cross national boundaries. The Migratory Species Convention, which is potentially a strong one, is therefore of great importance. Governments should join the convention without delay and national and international organizations should assist them to implement it.

Development of New Law

Because of their force, strong international agreements are very important for the implementation of the World Conservation Strategy. The more effective conventions need the constant, vigorous support of governments, non governmental organizations, and international organizations; and the weaker ones need strengthening. Action should not be restricted to conventions, however. Attention should also be paid to some of the other ways in which international law develops: for example, the exploration of new concepts such as the application of environmental assessment procedures at the international level; and the elaboration of "soft law" such as the Declaration of the United Nations Conference on the Human Environment. Although soft law is unenforceable it is of great value because it provides a set of generally agreed standards of international behaviour and paves the way for the codification of such standards in a more binding form.

International Assistance

The funds spent by multilateral and bilateral development assistance agencies (more than $27,000 million in 1976) can do a great deal towards restoring the environment, tackling environmentally induced poverty, and enabling countries to make the best use of their resources, if the projects they support are environmentally sound. It is recommended that such agencies make every effort to:

— direct funds to reforestation, the restoration of degraded environments, and the protection of watersheds, of mangroves and other critical habitats for marine resources, and of genetic resources essential for development;

— assess all projects for their ecological implications and ensure that they are ecologically sound; assist governments to design ecologically appropriate policies and to establish and maintain effective conservation infrastructures.

Assistance should be made available to enable requesting nations to develop the capacity to carry out national conservation strategies, ecosystem evaluations and environmental as sessments, and to implement cross-sectoral conservation policies through appropriate legislation, training and organization. Development agencies should assist governments on request to establish the laws, institutions and procedures to enable them to conserve their country's living resources. Developing industries based wholly or partly on living resources should be encouraged and equipped to ensure that the resources are exploited sustainably and that the genetic diversity on which ultimately they depend is preserved. Development assistance agencies have a special responsibility to help)—through the provision of appropriate advice and technical assistance—the recipient nation ensure that financial assistance makes the best use of the living resources it is likely to affect, and should endeavour to ensure that:

— the proposed development is compatible with the recipient country's national conservation policy and strategy (if they exist);

— the proposed development is the most appropriate response to the capabilities of the ecosystems concerned;

— an environmental assessment is carried out.

Nations lacking the capacity to prepare or carry out national conservation strategies, ecosystem evaluations or environmental assessments or lacking adequate conservation laws, means of enforcing them, or organizations to effect the full range of required conservation measures, should seek appropriate multilateral or bilateral assistance.

Tropical Forests and Drylands

Tropical Forests

Tropical forests are an important renewable resource, acting as a reservoir of genetic diversity, yielding a continual supply of forest products if managed sustainably, helping to regenerate soils and protect them from erosion, protecting areas downstream from floods and siltation, buffering variations in climate, and providing recreation and tourism. If tropical forests are exploited—as generally they are—with scant regard for their ecological characteristics, the resource cannot renew itself.

Hence while temperate forests seem to have reached a state of equilibrium, tropical forests are contracting rapidly as a result of expanding shifting agriculture, spontaneous settlement, planned colonization, clearance for plantations and ranching, cutting for fuel, and logging. It has been estimated that tropical rain forests are being felled and burned at the rate of 110,000 km a year; at this rate *all* of this forest type will have disappeared within 85 years.

Tropical forests are not uniform, however; nor is their rate of disappearance. The most valuable, and the richest in species, are lowland rain forests—and these are being destroyed, at a much faster rate. Many, like the lowland forests of Southeast Asia, are believed unlikely to survive to the end of this century.

International action to conserve tropical forests is required most urgently in West and East Africa, South and Southeast Asia, Central America and Mexico, and parts of South America. In West and East Africa the most important needs are:

— establishment of fuelwood plantations and of industrial plantations (existing and planned plantations will not even approach compensating the region for the projected loss of its natural forests); in those countries that have established national parks or nature reserves to protect genetic resources, strengthening of such parks or reserves, notably by allowing local people to benefit from buffer zones and by making surrounding areas priorities for rural development; in those countries that have not established parks or reserves, or where they are insufficient, identification of areas of greatest genetic importance and where human pressure is least, and establishment of parks and reserves in those areas;
— strengthening of administrations responsible for protection and management of natural forests;
— reforestation including well designed and strategically located forests to meet immediate needs for raw materials, to serve as model examples for the development of neighbouring areas, and to replace areas of forest already destroyed.

Madagascar, the relict forests of Ethiopia and East Africa's mountains, and Ivory Coast are Africa's priority areas for genetic resource preservation through the establishment of protected areas; but this will not be possible without greatly accelerated rural development outside the protected areas. Hence what is needed is a combined rural development/protected area programme.

Similar measures are required in Asia, where priority should be given to the exceptionally rich lowland (below 300 metres) dipterocarp forests of Borneo, peninsular Malaysia, Sumatra and the Philippines, and in the Americas, where priority should be given to the western Amazon basin, the Pacific coast areas of Colombia and Ecuador, and coastal and southeastern Brazil. The softwood (coniferous) forests of south-eastern Brazil are projected to decline from 5.8 million hectares in 1975 to 0.8 million hectares in the year 2000 (a loss of 86 %). Softwood forest depletion is also serious in Central America (a loss of 20%) and the Caribbean (a loss of 22%). Since the pine species of these regions provide the genetic raw materials for the forestation programmes of many other tropical countries, urgent conservation measures are required.

Regions where depletion pressures are less acute also need action, although its emphasis will be somewhat different. In such regions there is time to establish networks of protected areas designed systematically to safeguard a comprehensive range of the genetic diversity of tropical forests, especially of the tropical rain forests. These regions are: the Caribbean, Central Africa, developing Oceania, and parts of tropical South America and of continental Southeast Asia. In these regions also there should be greater scope for experimental research and management to develop productive, sustainable systems for the utilization of tropical forests.

In all regions there is an acute need to: protect areas of unusual genetic diversity; promote rural development based on production systems that enable a high proportion of forest cover to be retained; develop systems of commercial exploitation that utilize products other than timber (for example, drugs, gums and resins, natural silk); ensure that felling and planting programmes are sustainable. The cooperation of developed countries is required to make sure that their demand for tropical forest products does not exceed the capacity of tropical forest countries to supply them non-destructively.

Efforts should be continued: to develop more efficient means of using tropical woods and other forest products ; to reduce waste and incidental destruction during exploitation; to stabilize markets in tropical timbers; to develop plantations (using areas already cleared) to meet forthcoming world needs for wood products. Suitable case studies are required to demonstrate the economic and other benefits to be derived from well planned and zoned multipurpose use of tropical rain forests for the production of timber and other forest produce, wildlife management, recreation, scientific use, and so on. A code of practice should be drawn up jointly by the forestry departments of the main tropical forest countries and representatives of the international timber trade to govern the granting of concessions and the conduct of all extraction operations.

Drylands

Drylands—areas where rainfall is low and evaporation and transpiration are high—cover about a third (50 million km) of the earth's land surface. Unless used with care and skill, they are extremely prone to desertification—the gradual destruction or reduction of the land's capacity for plant and animal production. Desertification threatens the future of some 628 million people, of whom about 78 million are now directly affected by a drop in productivity of the land on which they depend. Regions already in the grip of desertification or at high to very high risk cover 20 million km—an area twice the size of Canada. The problem areas are arid and semi-arid lands, 95% of which are threatened with desertification (compared with 28% of sub-humid lands).

Desertification is a response to the inherent vulnerability of the land and the pressure of human activities. Pressure of human numbers and numbers of livestock, together with unwise development projects, the extension of rainfed agriculture into unsuitable areas, inadequate management of irrigated agriculture, overgrazing, and overcollection of firewood, have already degraded vast areas and caused great human suffering. These pressures continue.

The United Nations Conference on Desertification (Nairobi, 29 August-9 September 1977) synthesized into a Plan of Action a wide range of activities addressing the complex biological, social, economic and political factors involved, based on the development of proper land use, including conservation and enhancement of living resources and water resources. It envisaged the successful halting of desertification throughout the world by the end of this century through national programmes assisted by intergovernmental and nongovernmental organizations coordinated by UNEP. The funds necessary were seen as coming from enhanced levels of multilateral and bilateral assistance reinforced if possible by arrangements such as a special anti-desertification fund and an international taxation scheme. There is growing concern at the lack of implementation of this Plan of Action.

The problem, therefore, is not one of not knowing what to do but of getting agreed action done. National and international organizations should make special efforts to persuade multilateral and bilateral assistance agencies to support implementation of the Plan of Action. In countries where centuries of intensive human use have devastated the vegetation of large areas of dryland, there is much to rehabilitate but little left to preserve in the unexploited state. Emphasis should be on rehabilitation in areas of high human and animal population densities. Because of the high densities, the supply of alternative food, fuel, and employment, however difficult, is mandatory. Some immediate reduction of firewood pressure is possible by use of more efficient wood stoves and better insulation of dwellings. The areas concerned are marked on the World Map of Desertification. In other countries, the emphasis should be on protecting some of the many remaining unexploited areas. Then there are still other countries which have both large areas of degraded dryland and large areas which have yet to be intensively exploited. Emphasis in these should be on both rehabilitation and protection.

Several international activities already under way should be encouraged and expanded:

- *Promotion of schemes for rehabilitation of natural vegetation,* including the selection of prime areas for demonstration projects such as the UNEP/Unesco Integrated Project on Arid Lands, and similar work by Iran and the USSR. Enlightened programmes such as these should be encouraged and expanded, and the results monitored, widely publicized and used for demonstration. Particular encouragement should be given to schemes that reflect concern for biological diversity.
- *Identification and promotion of prime areas for protection,* including support for the Unesco/UNEP project to establish a comprehensive series of biosphere reserves in arid and semi-arid regions.
- *Encouragement and support of research in the ecosystems of arid and semi-arid lands* with the object of improving systems of management to restore the potential of these lands and enable them to be used in a sustainable manner.

Global Programme for the Protection of Genetic Resource Areas

Programmes to preserve genetic resources tend to be conducted along narrow sectoral lines: crops, forage plants, timber trees, livestock, microorganisms, animals for aquaculture, wildlife, and so on. This is justifiable in the case of off site preservation since each sector does have different requirements and therefore needs different collecting programmes. However, it is not the best approach to the on site protection of wild species. Since only a small proportion of the earth's surface is likely to be available for long term protection, each protected area should be so sited and managed as to protect as much genetic material as possible. This requires coordination among all the sectors concerned, as well as accurate knowledge of the locations of the species, subspecies and varieties most in need of protection. As the basis of continuing evolution, wild gene pools are the common heritage of mankind. They therefore need an international cross-sectoral programme for their protection on site.

It is recommended that each sector identify those species, subspecies and varieties most in need of on site protection ; their distribution; and those areas where their distribution overlaps. Such areas of overlap—or concentrations of priority species, subspecies and varieties—should be selected as priority areas for protection. In some cases the areas may be protected already. In all cases, governments should be encouraged to commit themselves to safeguarding the protected areas; and those commercial sectors which use the resources concerned should be persuaded to contribute to the financial costs of protection.

There are essentially three types of concentration of genetic resources: concentrations of wild and weedy relatives of species of economic or useful value; concentrations of threatened species (regardless of economic or useful value); ecosystems of exceptional diversity. In addition, ecosystems (regardless of their diversity) that are unrepresented or poorly represented in protected areas should also be given priority.

Concentrations of Economic or Useful Varieties

Global priorities for crops have been established by the International Board for Plant Genetic Resources (IBPGR) which has also determined regional priorities for South Asia and Southeast Asia. Although on site protection is seldom specified, several regional IBPGR reports recommend it and it is clearly essential for the wild relatives of many crops, these often being the major source of pest and disease resistance and other important adaptations. On site protection of such species and varieties preserves their ability to continue adapting to any change in the populations of the pathogens affecting them.

The IBPGR has identified ten high priority regions for the collection and preservation of crop genetic resources: Mediterranean, West Africa, Ethiopia, Southwest Asia, South Asia, Southeast Asia, Central Asia, Mexico and Central America, the Andean zone, and Brazil. These, therefore, are also the priority regions for selecting protected areas. However, more detailed analysis of the distribution of the wild relatives of priority crops is needed before this can be done.

Priorities for trees have been proposed by the FAO Panel of Experts on Forest Gene Resources. More than 130 tree species have been identified as in need of on site protection, most of them in six regions: Africa, Southeast Asia, Australia, Mexico, Caribbean and Central America, and Eastern USA and Canada. Priorities have not been established for aquatic animals, livestock, or microorganisms. In the case of livestock, however, traditional domesticated breeds are likely to remain more important for breeding than wild relatives; and microorganisms are unlikely to need special measures for on site protection, off site preservation in culture collections being entirely adequate.

Concentrations of Threatened Species

The most serious threat to plant species is habitat destruction. Plants threatened in this way are concentrated in the following ecosystem groups: islands, especially oceanic islands in the tropics and subtropics; tropical rain forests; drylands; Mediterranean-type ecosystems; wetlands, especially freshwater ones (notably in Europe).

Similarly, the three most serious threats to vertebrates are habitat destruction, overexploitation, and the effects of introduced species (respectively affecting 67%, 37% and 19% of threatened vertebrate species). Concentrations of species threatened by habitat destruction closely overlap concentrations of species threatened by the effects of introductions. Vertebrates threatened by habitat destruction are concentrated in the following ecosystem groups: fresh waters, especially in North America and in Africa; islands, especially in the tropics; tropical rain forests; wetlands; tropical dry or deciduous forests.

More than half the vertebrate species threatened by habitat destruction and 70% of the vertebrate species threatened by the effects of introduced species are concentrated in ten areas: fresh waters in North America and Mexico, West and Central Africa, Southern Africa; islands in Caribbean, Western Indian Ocean, South Pacific, Hawaiian Islands; tropical forests in Southeast Asia, Madagascar, South America.

Ecosystems of Exceptional Diversity

Terrestrial ecosystems of exceptional diversity include tropical rain forests (especially those of peninsular Malaysia, Borneo, Sulawesi, Sumatra, Philippines, New Guinea, Central and South America, and Madagascar), the tropical dry forests of Madagascar, the Mediterranean-type ecosystems of South Africa and Western Australia, and very rich island systems such as New Caledonia and the Hawaiian Islands. Marine ecosystems of exceptional diversity include the coral ecosystems of the Indo-Malay archipelago, the Western Pacific, the Northeast Pacific, the South-east Pacific, the Caribbean, and the Southeast and Southwest Atlantic. In addition, the Gulf of California, Gulf of Mexico, Red Sea, Sea of Okhotsk, Sea of Japan, and China Sea are important for their unique species. Exceptionally diverse freshwater ecosystems include the rivers of Amazonia, the rivers of West and Central Africa, the lakes of East and Central Africa, the Caspian and Aral Seas and Lake Baikal in the USSR, the Mississippi drainage of North America, the rivers of India, and the fresh waters of Borneo, Java and Sumatra.

Ecosystems that are Unrepresented or Poorly Represented in Protected Areas

Review of the distribution of protected areas of the land shows that 35 of the 193 biogeographical provinces listed have no national parks or equivalent reserves and a further 38, while having at least one park or reserve, are inadequately covered. Representative samples of the ecosystems in these provinces should be given protection as soon as possible. In addition, *all* the other provinces need to be examined for the quality of their coverage. In many of them only a small proportion of ecosystem-types are covered by protected areas, and the protection of many of them is inadequate. Special attention should be paid to concentrations of unique species within each province and to areas of exceptionally high diversity—notably low land forests (those below 300 metres intropical rain forests), tropical and subtropical cloud forests, and isolated mountains. It is not possible to indicate priorities for marine or freshwater ecosystems ; but these are much more poorly represented in protected areas than are terrestrial ones and deserve a correspondingly higher priority.

Similarly, although the establishment of an international network of biosphere reserves has progressed well, there is still a long way to go before the network is complete. The 162 biosphere reserves so far designated occur in 76 of the 193 biogeographical provinces. Some 50 biosphere reserves have been set up for mixed mountain and highland systems; about 30 are representative of temperate broadleaf forests; and only a dozen exist for tropical rain forests. Nevertheless, the most critical gaps, with fewer than six biosphere reserves, are in the following ecosystem groups: subtropical rainforests; temperate needleleaf forests; cold winter deserts and semi-deserts; tropical grasslands.and savannas; temperate grasslands; river and lake systems.

Financing the Global Programme

Protecting genetic resources is an international responsibility and the costs and benefits of doing so should be shared equitably. Many countries particularly rich in genetic resources are developing ones that can ill-afford to bear alone the burden of their on site protection. An international mechanism is needed by which those countries with an especially heavy responsibility can be compensated. The assistance provided by the IBPGR in the case of crop genetic resources is a start in this direction; but all types of genetic resources should be covered, and contributions should come not only from governments and international organizations but also from the commercial enterprises which benefit directly from living resources.

Commercial participation should be regarded not as conferring special rights of use but in recognition of shared responsibility. Unfortunately, commercial plant breeders and seed suppliers increasingly are patenting varieties and demanding royalties on their use even though the varieties are as much products of freely obtained genetic diversity as they are of commercial investment. As a result, many countries may have to pay twice over for genetic material—once for the new variety and once for protecting the material from which it is derived. Plant breeders' rights and the standardization of plant varieties should be so limited that neither has the effect of restricting the free exchange and use of genetic materials or of reducing genetic diversity.

Commercial Sponsorship of on Site Genetic Resource Protection

Industries and other commercial enterprises based on, or regularly using, particular plant or animal species should sponsor the establishment and maintenance of protected areas for the preservation of the relevant species, its relatives and varieties. Such areas should be regarded as crop and commodity banks on which the industrial sector concerned can draw for the development of new strains of plant or animal with whatever properties of productivity, pest or disease resistance, responsiveness to different soils and climates, nutritional quality, and so on, may be required.

Similarly, industries and other commercial enterprises that depend on naturally occurring chemical compounds either for raw materials or for product ideas should sponsor the establishment and maintenance of protected areas for the preservation of representative samples of ecosystem types, unique ecosystems, the habitats of unique and of threatened species, and other ecosystems essential for the preservation of genetic diversity. Such areas should be regarded as potential product banks on which the industrial sector concerned (for example, pharmaceuticals) could draw for the development of new or improved products.

Many businesses benefit from wild plants and animals, often in ways that may easily be overlooked. For example, alginate compounds from brown seaweeds are used in shampoos, soaps, cosmetics, paints, dyes, fire-extinguishing foams, building materials (insulation products, sealing compounds, artificial wood), and the lubricants and coolants used in drilling for oil. Every industry, therefore, should analyze its resource base to determine what living resources it uses and for what purpose, and the extent to which each resource's combination of desired properties, cost and availability is unique to that resource. Each industry should then work with the governments and the other commercial sectors concerned to ensure that the particular plants and animals are exploited sustain-ably, that their genetic diversity is preserved, and that the ecological processes of which they are part are maintained. Such measures would go some way towards ensuring the quality and availability, at reasonable prices, of valuable raw materials.

The Global Commons

A commons is a tract of land or water owned or used jointly by the members of a community. The global commons includes those parts of the earth's surface beyond national jurisdictions—notably the open ocean and the living resources found there—or held in common—notably the atmosphere. The only land mass that may be regarded as part of the global commons is Antarctica, although several countries have claimed parts of it (the claims are currently frozen under the Antarctic Treaty)!.

The Open Ocean and its Living Resources

Much of the open ocean remains frontier country in which people can exploit living resources as they please as long as they have the technology to do so. While the open ocean is not as biologically rich as continental shelf areas, it includes unique ecosystems and provides some (and in some cases all) of the critical habitats of several culturally and economically important groups of animals, notably whales and tunas. Species that are confined to the open ocean should be regarded as the common resource of all humanity; species that move between the open ocean and waters under national jurisdiction are shared resources. Special provision for the conservation of both groups of species is therefore needed, but no satisfactory mechanism exists.

Living resource exploitation in the open ocean is regulated only in the case of two groups of species: tunas and whales. There is no protection of the habitats of open ocean species and hitherto none has been needed. However, with the advent of deep sea mining and the increasingly intensive use of ocean space generally, protection is now required. The designation by the International Whaling Commission (IWC) of an Indian Ocean Sanctuary in which all commercial whaling is prohibited, while an encouraging step forward, needs to be matched by international measures to protect the habitats of the whales, dolphins and porpoises in that area.

Means of ensuring the conservation of open ocean species and ecosystems— particularly ecosystems supporting the feeding grounds of whales, salmon, and so on, the spawning grounds of tuna, unique areas and areas of unusual species diversity—should be devised, promoted and adopted. An appropriate international organization should prepare a discussion document, possibly as a prelude to an expert consultation, on priority species and ecosystems and on ways and means of conserving them. These might include the introduction of more effective measures to regulate exploitation and the establishment of sanctuaries where the habitats of cetaceans and other marine creatures are protected and exploitation is prohibited.

The International Whaling Commission has imposed a moratorium on the taking, killing, or treating of whales, except minke whales, by factory ships or whale catchers attached to factory ships. The moratorium should be extended to all commercial whaling until: the consequences for the ecosystems concerned of removing large portions of the whales' populations, and such populations' capacity for recovery, can be predicted; permitted levels of exploitation are safe, and an effective mechanism exists for detecting and correcting mistakes in the management of any stock; member nations of the IWC are no longer purchasing whale products from, or transferring whaling technology and equipment to, or otherwise supporting, non-member nations, or pirate whaling ships.

The dumping of wastes at sea is regulated by the Convention on the Prevention of Marine Pollution by Dumping of Wastes and Other Matter and by regional conventions for which it provides the framework. Those states not yet party to this convention are urged to join it. It is also important to control the effects of deep sea mining (including oil exploitation); for this to be done it is necessary to learn what those effects are.

Accordingly, an internationally agreed area should be established, in which deep sea mining is prohibited, as a base line area to aid the long term evaluation of the effects of deep sea mining.

In addition, all nations engaged in, or considering, deep sea mining activities—or any other activities with currently unpredictable effects on open ocean ecosystems—should:

— *precede commercial mining operations or similar activities by commissioning a comprehensive ecological survey* to determine the impact of such activities;

— *designate appropriate areas of the deep sea bed as baseline reference and resource zones* in which no mining or other significant disturbance will be allowed, ensuring that the size and shape of each area is such that its stability will be maintained; *establish guidelines for scientific research* to ensure minimum disruption of the natural state of such areas, and provide for full exchange of information on the results of research.

The Atmosphere and Climate

The behaviour of the atmosphere, like that of the ocean, is indifferent to political boundaries. Impacts on the atmosphere in one country can affect the living resources of other countries, both directly and by altering climate. Such impacts are increasing. Acid rain caused by excessive emissions of sulphur dioxide, mainly in Europe and North America, has reduced the productivity of many lakes, rivers and forests in countries other than the sources of the pollution. The accumulation in the atmosphere of other gases—such as carbon tetrachloride and methvlchloroform (used in industrial solvents), nitrous oxides from the decomposition of nitrogen compounds, the chlorofluoromethanes (used in refrigerators, air conditioners and aerosol sprays) and carbon dioxide—is potentially a still more serious problem because of its possible effects on climate. For example, it has been estimated that if chlorofluoromethanes continue to be released at current rates, the ozone layer could be reduced by up to 15% by the middle of the next century, which in turn could impair human health and the productivity of the biosphere.

Major alterations of the land surface—such as forest clearance, massive water impoundments and irrigation sys-terns, and the expansion of urban areas—also have the potential to alter local or regional climates, for example by changing heat and moisture exchange between the surface and the atmosphere. The most acute climatic problem, however, is carbon dioxide accumulation, as a result of the burning of fossil fuels, deforestation and changes in land use.

At present rates of increase, the atmospheric concentration-of carbon dioxide may produce a significant warming of the lower atmosphere before the middle of the next century, particularly in the polar regions. This wanning would probably change temperature patterns throughout most of the world, benefitting some regions and damaging others, possibly severely.

Since it would be necessary to redirect many aspects of the world economy, including energy production and agriculture, to halt or reduce further at mospheric carbon dioxide accumulation, accelerated research at the national and international levels is required to determine more precisely the likely climatic and other effects and their socio econoic consequences. There is also a general need for better climatic data, for clarification of the relative roles of human and natural influences on climate, and for improved understanding of the impact of climate change on human activities.

Research into all these issues is the main thrust of the World Climate Programme sponsored by the World Meteorological Organization, and this programme deserves the strongest support of all nations. In addition, the immediate problem of acid rain requires not merely continued research but a reduction by European and North American countries of sulphur dioxide emissions. It is most important that states join and implement the Convention on Long-range Transboundary Air Pollution.

Antarctica and the Southern Ocean

Antarctica and the Southern Ocean are defined as all land and sea south of the Antarctic Convergence. Much of this area—that is the entire area south of 60° S latitude except for the "high seas"—is under the nominal control of the 13 parties to the Antarctic Treaty.

Under this treaty, Antarctica may be used only for peaceful purposes—principally scientific research. Conservation of the living resources of the land is provided for by the "Agreed Measures for the Conservation of Antarctic Fauna and Flora" which are excellent but have not yet been ratified.

Currently the potential of krill *(Euphausia superba),* a tiny shrimp-like creature, swarming in huge quantities in the Southern Ocean, is attracting a great deal of interest: it is said that the catch could rise from about 50,000 tonnes in 1977/1978 to 60 million tonnes or more—thereby doubling the world's current annual fish catch. However, krill are the major food of five species of great whales, including the endangered blue whale and humpback whale, and are also important for three species of seals, many seabird species and several species of fish. Unless krill harvesting is extremely carefully and conservatively regulated, its effects on other Southern Ocean species could be devastating. A convention to regulate the taking of the living resources of the Southern Ocean is being negotiated, and is expected to be followed by a regime for mining and oil exploitation.

Any regime for the exploitation of the living marine resources of the Southern Ocean should so regulate the krill fishery as to prevent: irreversible changes in the populations of krill; irreversible changes in the populations of the baleen whales and those seal, fish and bird species which feed on krill, as well as in the Southern Ocean ecosystem as a whole; overcapitalization of krill fishing fleets, which could make it more difficult to agree on a reduction of the krill take should this prove necessary, and could have severe impacts on fisheries outside the Southern Ocean, due to the need to redeploy the krill fleets during the Antarctic winter. An independent observer system should be provided for in such regulations.

The Antarctic Treaty powers and nations fishing or intending to fish the Southern Ocean should exercise extreme restraint on catch levels until understanding of this uniquely productive ecosystem improves. All harvesting should be on an experimental basis as part of a scientific research programme to improve knowledge of krill and of the Southern Ocean as a whole. Baseline areas where no krill or other living or non-living resources may be taken should be set aside and given complete protection, so that impacts outside can be monitored and evaluated correctly. The dimension and location of these areas should be established according to the best available knowledge of the ecosystems concerned. Current research efforts should be strongly supported; and the collection, analysis and dissemination of biological information should be mandatory. An International Decade of Southern Ocean Research, focussing particularly on ecological processes, should be initiated as a matter of urgency. Investigation of the possible environmental impacts of tourism, scientific research, mining and oil exploitation, and so on, should be continued. Since oil degrades, extremely slowly in conditions such as those of Antarctica and since operating hazards are very high, the feasibility of oil exploration and exploitation in particular should be approached with the utmost caution.

Regional Strategies for International River Basin and Sea

The purpose of regional strategies is to stimulate national action where it is most needed, to help solve common problems, and in particular to advance the conservation of shared living resources. Each regional strategy, which should be prepared along the lines of national strategies, should aim for at least four products: agreements on the joint conservation of shared living resources; model examples of how common problems can be tackled successfully; joint organizations where appropriate and where more cost-effective than several national organizations (for example, for training, for research and monitoring, or for the management of shared living resources); improved information for national decision making.

Each "region" should be an ecological unit in which by definition many of the living resources will be shared; Obvious examples, and priority candidates for regional strategies, are international river basins and seas.

Soil and water conservation and the conservation of marine living resources not only require a cross-sectoral approach at the national level, they frequently demand international cooperation as well. This is certainly the case with international river basins and seas. International river basins are drainage basins or catchment areas shared by two or more states and communicating directly with the sea or inland lakes. International seas are either semi-enclosed seas shared by two or more states or more open seas in which areas under the jurisdiction of two or more states are closely linked by currents or by animal migrations. Both in international river basins and in international seas the living resources of one country are likely to be affected by events in another—such as pollution, alteration of habitats, or overfishing. Hence international cooperation is generally necessary for pollution control and for rational utilization of resources. Cooperation also provides opportunities to improve efficiency and achieve economies through joint action as well as for international technical and financial assistance to support that action.

International River Basins

There are more than 200 international river basins: 57 in Africa, 48 in Europe, 40 in Asia, 36 in South America and 33 in North and Central America· They include river basins experiencing the most severe soil erosion in the world (the amount of sediment load in relation to the size of the drainage basin is a measure of the intensity of erosion).

Table 1. International River Basins

River	***Drainage basin thousand***	***Average annual suspended load***	
million km	***tonnes/tonnes***	***km***	
Huang (Yellow)	673	1,887	2,804
Ganges*	956	1,451	1,518
Brahmaputra*	666	726	1,090
Yangtze	1,942	499	257
Indus*	969	435	449
Ching	57	408	7,158
Amazon*	5,776	363	63
Mississippi	3,222	312	97
Irrawaddy*	430	299	695
Missouri	1,370	218	159
Lo	26	190	7,308
Kosi	62	172	2,774
Mekong*	795	170	214
Colorado*	637	135	212
Red*	119	130	1,092
Nile*	2,978	111	37

* International river basin.

Major Rivers of the World Ranked by Sediment Load.

Joint use of watercourses has always depended on cooperation among the riparian states, and some of the oldest international organizations were created to manage river navigation on the Rhine and the Danube. The use of international inland waters has steadily expanded: new industrial, urban and agricultural demands on water quantity have risen more or less simultaneously with a dramatic decline in water quality in most international basins. Forest clearance, hydroelectric installations, irrigation and water supply works and pollution in one country can rob another of water, increase its costs of making water suitable for different uses, and destroy, degrade or deplete its valuable ecosystems and species.

Failure to reconcile the competing interests of upstream and downstream users has generated considerable political friction in many parts of the world. Where traditional interstate basin commissions exist, they are often ill-adapted to the new challenge of water conservation and integrated environmental management. There have been some no-table regional improvements, however— such as the Danube fisheries conservation agreement, the Great Lakes water quality agreement, the Rhine salinity and chemical pollution agreements, and developments in the Mekong Basin Commission and the Lake Chad Basin Commission.

To evaluate these experiences, with a view to their adaptation in other regions, an appropriate international or ganization should undertake a review of the conservation needs and problems of international river basins, as a prelude to joint research and action plans by the riparian countries concerned, possibly along the lines of UNEP's regional seas programme. Priority should be given to international river basins scheduled for major development or subject to severe erosion. Regions most susceptible to erosion are those in the tropics receiving medium to high rain fall; there is also a positive correlation between heavy sediment loads and close proximity of mountains to the sea.

International Seas

Most large maritime nations and several smaller ones have extended their national jurisdictions by declaring Exclusive Economic Zones (EEZs) for 200 nautical miles from their shores. Others are likely to do the same. These moves mean that the international agreements covering most regional fisheries commissions must be renegotiated. Several have been already. There are signs, however, that ecological considerations are being given insufficient weight and that this opportunity will be missed.

The establishment of EEZs adds to the incentives for coastal states to protect the habitats critical for fisheries since they now control the fisheries—at least of non-migratory species—which the habitats support. By protecting the habitats, and by seeing to it that the fisheries themselves are exploited on a sustainable basis, they will assure both a regular high quality protein supply and often a substantial income. Many species, however, move between one EEZ and another, and between EEZs and the ocean beyond national jurisdiction. In addition, as the results of oil spills regularly demonstrate, currents carry pollutants from one EEZ to another. Therefore the need for international cooperation and for ecologically sound management remains unchanged.

New or improved bilateral and mul tilateral management agreements are needed to ensure that marine pollution is controlled and marine living resources exploited sustainably. Regional efforts to regulate pollution of the sea—notably those being made by governments in cooperation with UNEP's regional seas programme—should be intensified, and similar efforts should be initiated wher ever groups of nations share common bodies of water. Experience with UNEP's regional seas programme has shown that there is great scope for regional agreements, elaborated by specific technical protocols, and backed up if neces sary by the establishment of regional organizations.

Regional strategies should pay particular attention to: the status of fisheries and other living resources and measures to ensure they are utilized sustainably; the protection and main tenance of the critical habitats (feeding, breeding, nursery, and resting areas) of economically or culturally important species and of threatened or unique species; the preservation of genetically rich areas such as coral reefs; the pro tection and maintenance of the support systems of critical habitats and of ge netically rich areas; measures to control pollution and as far as possible to pre vent accidents such as oil spills; provi sion for a rapid and effective response to such accidents.

The causes, magnitude and consequences of environmental problems should be evaluated; critical habitats, genetically rich areas and their support systems should be mapped (showing, where known, the rough tim ing and periodicity of the processes involved and the extent to which critical habitats have been observed to change with variations in climate and other environmental factors); and present and projected uses of and impacts on the ecosystems and species concerned should be analyzed so that compatibilities and conflicts may be revealed and decisions made accordingly.

Regions likely to benefit most from marine conservation strategies are: regions that depend heavily on marine living resources (whether for food or foreign exchange); regions in which countries have gained major fisheries (bigger than 50,000 tonnes per year) as a result of extending national jurisdictions to 200 nautical miles from shore; or regions where international conservation programmes have already begun.

Because the Arctic environment takes so long to recover from damage, the Arctic should be considered a priority sea. Within their Arctic territories the Arctic nations should systematically map critical ecological areas (terrestrial as well as marine), draw up guidelines for their long term management, and establish a network of protected areas to safeguard representative, unique and critical ecosystems.

Since various conservation problems in the Arctic relate to areas or populations beyond national jurisdiction or which are of common concern to two or more of the Arctic nations, a meeting to identify and discuss such problems would probably facilitate conservation in the region. Among items of common concern are: measures (including joint research) to improve protection of migratory species breeding within the Arctic and wintering inside or outside the region; studies of the impact of fisheries and other economic activities in the northern seas on ecosystems and non-target species; the possibility of developing agreements among the Arctic nations on the conservation of the region's vital biological resources, based on the principles and experience of the Agreement on Conservation of Polar Bears.

Shared Resources

Shared resources are -defined here as ecosystems and species shared by two or more states (including species that move between one national jurisdiction and another) and ecosystems and species that depend on or are affected by events in another. They include ecosystems and species of international river basins and many coastal ecosystems and associated fisheries ; and they also include migratory species. It is strongly urged that all states observe the Draft Principles of Conduct in the Field of the Environment for the Guidance of States in the Conservation and Harmonious Utilization of Natural Resources Shared by Two or More States, prepared by UNEP and recommended by its Governing Council to the General Assembly of the United Nations for adoption. These principles stress the need for states to:

— cooperate in controlling, preventing, reducing or eliminating adverse environmental effects that may arise from the utilization of shared natural resources; avoid environmental damage that could have repercussions on the utilization of the resource by another sharing state;

— make environmental impact assessments before engaging in any activity with respect to a shared natural resource that may significantly affect the resource or the environment of another sharing state; give other sharing states in advance the details of any plans to begin or change the conservation or utilization of a shared natural resource, and consult with them and provide additional pertinent information on request;

— engage in joint scientific studies and assessments;

— compensate for damage to shared natural resources or arising out of the utilization of such resources; and provide persons in other states who have been or may be harmed by such damage with equivalent access to and treatment in the same administrative and judicial proceedings as are available to nationals.

Towards Sustainable Development

Development and conservation operate in the same global context, and the underlying problems that must be overcome if either is to be successful are identical.

Much habitat destruction and over-exploitation of living resources by individuals, communities and nations in the developing world is a response to relative poverty, caused or exacerbated by a combination of human population growth and inequities within and among nations. Peasant communities, for example, may be forced to cultivate steep, unstable slopes because their growing numbers exceed the capacity of the land and because the fertile, easily managed valley bottoms have been taken over by large land-owners. Similarly, many developing countries have so few natural resources and operate under such unfavourable conditions of international trade that often they have very little choice but to exploit forests, fisheries and other living resources unsustainably. In many parts of the world population pressures are making demands on resources beyond the capacity of those resources to sustain. Every country should have a conscious population policy to avoid as far as possible the spread of such situations, and eventually to achieve a balance between numbers and environment. At the same time, it is essential that the affluent constrain their demands on resources, and preferably reduce them, shifting some of their wealth to assisting the deprived. To a significant extent the survival and future of the poor depends on conservation and sharing by the rich.

During the 1980s, the Third United Nations Development Decade, the efforts of the international community to remove the main obstacles to development and to raise the living standards of the poor in a sustained and rapid manner will focus on the new International Development Strategy. The ultimate aims of this strategy are "(a) to redress the inequities in the relations between richer and poorer nations; (b) to establish a more dynamic, more stable and less vulnerable world economy, in which all countries have opportunities to participate on a fuller and more equal basis; (c) to stimulate accelerated economic growth in the poorer countries of the world; and (d) to reduce and eventually overcome the worst aspects of poverty by improving the lot of the hundreds of millions of people now living in abject poverty and despair".

The lack of progress so far in achiev ing these aims thwarts conservation as much as it does development. Hence it is as necessary for conservation as it is for development that during the 1980s:

a. trade be liberalized, including the removal of all trade barriers to goods from developing countries;

b. the flow of finance and development assistance be increased, including as a minimum the renewal of the objective of 0.7% of the gross national product of developed countries as official development assistance;

c. the proportion of development assis tance going to low-income countries (those countries with per capita in comes of $300 or less— in which live two-thirds of the poor in developing countries) be increased to at least two-thirds and preferably to three- quarters;

d. the international monetary system be reformed;

e. a code of conduct for transnational companies be adopted;

f. there be much more rapid progress on disarmament (expenditure on arms and military activities currently absorbs $400,000 million a year);

g. economic and social growth be accel erated, especially in the poorest coun tries, ensuring that economic and social goals are mutually supporting, and emphasizing better health, better housing, and higher educational levels and skills.

Achievement of equitable, sustain able development requires implementa tion not only of the measures indicated above but also of the World Conservation Strategy. Accordingly it is strongly - recommended that the objectives of the World Conservation Strategy—the maintenance of essential ecological processes and life-support systems, the preservation of genetic diversity, and the sustainable utilization of species and ecosystems—be included in the new In ternational Development Strategy.

Living resource conservation is es sential for the achievement of several development targets, for example: in creased growth in food production; development and efficient expansion of environmentally benign forms of energy; more efficient use of raw materials; prevention and reduction of desertifi cation, of soil degradation and loss, and of living resource overexploitation; and attainment of an acceptable level of health for all. Conservation is entirely compatible with the growing demand for "people-centred" development, that achieves a wider distribution of benefits to whole populations (better nutrition, health, education, family welfare, fuller employment, greater income security, protection from environmental degrada tion) ; that makes fuller use of people's labour, capabilities, motivations and creativity; and is more sensitive to cul tural heritage.

Coordination and Follow-up

The organizations most involved in the preparation of the World Conservation Strategy (IUCN, UNEP, WWF, FAO and Unesco), recognize the need to carry out international action to implement the Strategy and to stimulate and support national action. For its part, IUCN will promote the implemen tation of the World Conservation Strategy, particularly of national strate gies and of action at the international level; will monitor implementation as closely as possible; will publish regular news of implementation and will issue a full progress report every three years.

The progress report will cover:

— what governments and organizations are doing to implement the Strategy;
— whether what they are doing is likely to alleviate the problem or achieve the objective concerned; in due course, the extent to which the three conservation objectives have been achieved.

The problems posed by the destruction, degradation and depletion of living resources are many and complicated. The resources available to tackle them are small and priorities for their use are not always determined with sufficient care. There is a need to deal with the causes of many of these problems rather than with the symptoms. There are many competent, interested organiza tions with seemingly divergent but basically compatible aims that would be better able to tackle the problems if they cooperated more along agreed lines. It is hoped that this Strategy will help governments, intergovernmental bodies, private organizations and individuals to cooperate with each other and jointly deploy the limited means available to much greater effect. If this is done, then the prospects for conservation—and for sustainable development—will be much enhanced.

References

Beard, C. *Wildlife Conservation and the Roots of Environmentalism. The Facts and Figures.* 1997.

IUCN-UNEP-WWF. *World Conservation Strategy: Living Resource Conservation for Sustainable Development.* 1980

Santos, M.A. *The Environmental Crisis.* Guides to Historic Events of the Twentieth Century, Greenwood Press, Westport, Connecticut. 1999.

Smout, T.C. *Nature Contested – Environmental History in Scotland and Northern England Since 1600.* Edinburgh University Press, Edinburgh. 2000.

Stamp, D. *Nature Conservation in Britain.* Collins New Naturalist, London. 1969.

WCED. *Our Common Future.* The World Commission on Environment and Development. Oxford University Press, Oxford, UK. 1987.

Bibliography

Agnoletti, M. (Ed.) *The Conservation of Cultural Landscapes*. CAB International, Wallingford, Oxon, UK.

Balachandran, Sarojini, ed., *Encyclopedia of Environmental Information Sources,* Gale Research Inc., Washington, DC, 1993

Barth, S., et al., *Exploring the Theory and Application of Ecosystem Management*, University of Michigan, Ann Arbor, MI, 1994.

Barutciski, M. International law and development-induced displacement and resettlement. *In:* De Wet, C.J. (ed.) *Development-induced displacement: Problems, policies and people.* Berghan Books, Oxford, UK and New York, NY, USA. 2006.

Beard, C. *Wildlife Conservation and the Roots of Environmentalism. The Facts and Figures.*

Beder, S. *Environmental principles and policies: An interdisciplinary introduction.* University of New South Wales Press and Earthscan, London, UK. 2006.

Birnie, P.W. and Boyle, A. *International law and the environment.* Oxford University Press. Oxford, UK and New York, NY, USA.2006.

Blackman, A., Can Voluntary Environmental Regulation Work in Developing Countries? Lessons from Case Studies. *Policy Studies Journal,* 2008. 36(1): p. 119-141.

Caputo, Darryl F. *Open Space Pays: The Socioenvironomics of Open Space Preservation,* New Jersey Conservation Foundation.

Cassese, A. Self-determination of peoples: *A legal reappraisal* (Hersch Lauterpacht Memorial Lectures), Cambridge University Press, Cambridge, UK.1999.

Chape, S. Spalding, M. Jenkins M. *The world's protected areas: status, values and prospects in the 21st century.* Univ de Castilla La Mancha, 2008.

Cutter, S.L. & Renwick, W.H. *Exploitation, Conservation, Preservation – A Geographic Perspective on Natural Resource Use.* Third edition. John Wiley & Sons Ltd., New York. 1999.

Dowie, M. Conservation refugees: When protecting nature means kicking people out. *Orion Online* November-December: 1–12. 2005.

Driesen, D.M. 'Thirty years of international environmental law: A retrospective and plea for reinvigoration', 30 Syracuse J. Int'l L. and Com. 353. 2003.

Fox, Thomas, *Urban Open Space: An Investment That Pays,* Neighborhood Open Space Coalition, New York, NY, 1990.

Frank, James E., *The Costs of Alternative Development Patterns: A Review of the Literature,* the Urban Land Institute, 1989.

Freedman, Bill, *Environmental Ecology: The Ecological Effects of Pollution, Disturbance, and Other Stresses,* Academic Press, San Diego, CA, 1995.

Gaston, K.J. (Ed.) *Biodiversity: A Biology of Numbers and Difference.* Blackwell Science, Oxford.

Harding, R., Ecologically sustainable development: origins, implementation and challenges. *Desalination*, 2006. 187(1-3): p. 229-239

Interagency Ecosystem Management Task Force, *The Ecosystem Approach: Healthy Ecosystems and Sustainable Economies,* Report of The Interagency Ecosystem Management Task Force, Volume 1, June 1995.

International Council on Human Rights Policy (ICHRP). *Climate change and human rights: A rough guide.* ICHRP, Versoix, Switzerland. 2008.

IUCN-UNEP-WWF. *World Conservation Strategy: Living Resource Conservation for Sustainable Development.* 1980

Jeffries, M.J. *Biodiversity and Conservation*. Routledge, London.

Karamanos, P., Voluntary Environmental Agreements: Evolution and Definition of a New Environmental Policy Approach. *Journal of Environmental Planning and Management,* 2001. 44(1): p. 67-67-84.

Kevin Bishop et al., *Protected For Ever? Factors Shaping the Future of Protected Areas Policy*, 12 Land Use Policy 291. 1995.

Lund, H. Gyde. *Definitions of Forest, Deforestation, Afforestation, and Reforestation.* Gainesville, VA: Forest Information Services. 2006.

Michael J. B. Green & James Paine, State of the World's Protected Areas at the End of the Twentieth Century, Paper presented at IUCN World Commission on Protected Areas Symposium on *Protected Areas in the 21st Century: From Islands to Networks,* Albany, Australia. Nov. 24–29, 1997.

Mitchell, R.B., International Environmental Agreements: A Survey of Their Features, Formation, and Effects. *Annual Review of Environment and Resources*, 2003. 28(1543-5938, 1543-5938): p. 429-429-461.

Pepper, D. *Modern Environmentalism*. Routledge, London. 1996.

Riley, D. & Young, A. *World Vegetation*. Cambridge University Press, Cambridge.

Runge, F. 'A global environmental organization (GEO) and the World Trading System,' *Journal of World Trade,* 35(4), 399-426. 2001.

Santos, M.A. *The Environmental Crisis*. Guides to Historic Events of the Twentieth Century, Greenwood Press, Westport, Connecticut. 1999.

Sensi, S. Human rights and the environment: A practical guide for environmental activists. *Policy Matters 15: Conservation and Human Rights.* CEESP/IUCN, CENESTA, Tehran, Iran. 2007.

Smout, T.C. *Nature Contested – Environmental History in Scotland and Northern England Since 1600*. Edinburgh University Press, Edinburgh. 2000.

Speth, J. (ed.), *Worlds Apart: Globalization and the Environment,* Washington, DC: Island Press. 2003.

Stamp, D. *Nature Conservation in Britain*. Collins New Naturalist, London. 1969.

The California Institute of Public Affairs (CIPA) (August 2001)."An ecosystem approach to natural resource conservation in California". *CIPA Publication No. 106*. InterEnvironment Institute. Retrieved 10 July 2012.

United Nations Development Programme (UNDP). *Applying a human rights-based approach to development cooperation and programming: A UNDP capacity development resource.* 2006.

WCED *Our Common Future*. The World Commission on Environment and Development. Oxford University Press, Oxford, UK.

Wilson, E.O., *Biodiversity,* National Academy Press, Washington, DC, 1994.

Witte, J., C. Streck, and T. Benner (eds), *Progress or Peril? Partnerships and Networks in Global Environmental Governance*. The Post-Johannesburg Agenda, Washington, DC/Berlin: Global Public Policy Institute.2003.

Wofford, C., 'A greener future at the WTO: the refinement of WTO jurisprudence on environmental exceptions to the GATT', Harvard Environmental Law Review, 24: 563. 2000.